DISTRIBUTION TH

Principles and Applicat

DISTRIBUTION THEORY

Principles and Applications

Edited by

Fozia Homa, PhD

Mukti Khetan, PhD

Mohd. Arshad, PhD

Pradeep Mishra, PhD

AAP | APPLE ACADEMIC PRESS

First edition published 2024

Apple Academic Press Inc.
1265 Goldenrod Circle, NE,
Palm Bay, FL 32905 USA

760 Laurentian Drive, Unit 19,
Burlington, ON L7N 0A4, CANADA

CRC Press
6000 Broken Sound Parkway NW,
Suite 300, Boca Raton, FL 33487-2742 USA

4 Park Square, Milton Park,
Abingdon, Oxon, OX14 4RN UK

Library and Archives Canada Cataloguing in Publication

Title: Distribution theory : principles and applications / edited by Fozia Homa, PhD, Mukti Khetan, PhD, Mohd. Arshad, PhD, Pradeep Mishra, PhD.
Other titles: Distribution theory (Palm Bay, Fla.)
Names: Homa, Fozia, editor. | Khetan, Mukti, editor. | Arshad, Mohd. (Lecturer in mathematics), editor. | Mishra, Pradeep (Professor of statistics), editor.
Description: First edition. | Includes bibliographical references and index.
Identifiers: Canadiana (print) 20230150616 | Canadiana (ebook) 20230150683 | ISBN 9781774912140 (hardcover) | ISBN 9781774912157 (softcover) | ISBN 9781003336303 (ebook)
Subjects: LCSH: Distribution (Probability theory)—Statistical methods—Data processing. | LCSH: Mathematical statistics—Data processing.
Classification: LCC QA273.6 .D57 2023 | DDC 519.2/4—dc23

Library of Congress Cataloging-in-Publication Data

CIP data on file with US Library of Congress

ISBN: 978-1-77491-214-0 (hbk)
ISBN: 978-1-77491-215-7 (pbk)
ISBN: 978-1-00333-630-3 (ebk)

About the Editors

Fozia Homa, PhD
Department of Statistics, Mathematics and Computer Application, Bihar Agricultural University, Sabour, Bhagalpur, Bihar, India

Fozia Homa, PhD, is an Assistant Professor and Scientist in the Department of Statistics, Mathematics and Computer Application, Bihar Agricultural University, Sabour, India, and author or co-author of several journal articles. Dr. Fozia has received numerous awards in recognition of her research and teaching achievements from several organizations of national and international repute. She has been conferred with the Young Scientist Award in the field of Mathematics and Statistics (2016) by Aufau Periodicals, Tamil Nadu, India. She was also awarded with a SP Dhall Distinguished Publication Award in Statistics (2015) by the Society for Advancement of Human and Nature, Himachal Pradesh, India; Young Scientist Award (2015) by the Venus International Foundation, Chennai, India; and Best Young Researcher Award (2015) by GRABS Educational Trust, Chennai, India. She has been an active member of the organizing committees of several national and international seminars, conferences, and summits. Dr. Homa acquired her BSc (Statistics Hons) and MSc (Statistics) degrees from Banaras Hindu University, Varanasi, Uttar Pradesh, India, and her PhD (Applied Statistics) with a specialization in sampling techniques from the Indian Institute of Technology (Indian School of Mines), Dhanbad, Jharkhand, India. She has received several grants from various funding agencies to carry out her research projects. Her areas of specialization include sample surveys, population studies, and mathematical modeling.

Mukti Khetan, PhD
Department of Mathematics, Amity University, Mumbai, Maharashtra, India

Mukti Khetan, PhD, is currently working as an Assistant Professor in the Department of Mathematics, Amity School of Applied Sciences, Amity University, Mumbai, Maharashtra, India. She formerly worked in the Department of Mathematics at IIT Bombay, Maharashtra, and in the Department of Statistics at Sambalpur University, Sambalpur. She completed her PhD in Statistics from IIT(ISM), Dhanbad, and her MSc degree in Statistics from Banaras Hindu University, Varanasi. She has published several research papers and book chapters in reputed national and international journals. She has supervised three MPhil students and has more than four years of teaching and research experience.

Mohd. Arshad, PhD
Department of Mathematics, IIT Indore, Madhya Pradesh, India

Mohd. Arshad, PhD, is currently working as an Assistant Professor in the Department of Mathematics, Indian Institute of Technology Indore, India. He has previously worked in the Department of Statistics and Operations Research at Aligarh Muslim University, India. After acquiring MSc degree in Statistics, Gold Medalist, from C.S.J.M. University, Kanpur, he completed his PhD in Statistics from IIT Kanpur. He has published several research papers in reputed international journals. He has supervised two PhD students and is presently supervising two PhD students. He is a member of the editorial boards of various journals and scientific societies. He is an author of a book titled, *Solutions to IIT JAM for Mathematical Statistics*. Dr. Arshad has been teaching undergraduate and postgraduate courses for more than seven years.

Pradeep Mishra, PhD
Department of Mathematics and Statistics, College of Agriculture, Jawaharlal Nehru Agricultural University, Jabalpur, Madhya Pradesh, India

Pradeep Mishra, PhD, is an Assistant Professor of statistics at the College of Agriculture at the Jawaharlal Nehru Agricultural University, Jabalpur, Madhya Pradesh, India. Dr. Mishra formerly served as a data management specialist at a private multinational company for several years. He specializes in the field of time series, design of experiments, and agriculture statistics. He has published more than 45 research articles in international and national journals. He has won several awards, including a Young Scientist Award in 2017 at the international conference Global Research Initiatives for Sustainable Agriculture and Allied Sciences (GRISAAS-2017), Best Doctoral Degree at the same meeting in 2018, and an award for a paper from the Society of Economics and Development in 2018, among other honors. Dr. Mishra earned his BSc in Agriculture from the College of Agriculture, Bilaspur, affiliated with the Indira Gandhi Krishi Vishwavidyalaya, Raipur (C.G.); his MSc in Agricultural Statistics from the College of Agriculture, Jawaharlal Nehru Agricultural University; and his PhD in Agriculture Statistics, specializing in modeling and forecasting of foods crops in India and their yield sustainability. During his PhD, he was selected for an INSPIRE Fellowship by the Department of Science and Technology of the Government of India due to his valuable research work.

Contents

Contributors

Mohd. Arshad
Department of Mathematics, Indian Institute of Technology Indore, Simrol, Indore, Madhya Pradesh, India

Suresh Badarla
Department of Mathematics, Amity School of Applied Sciences, Amity University, Mumbai, Maharashtra, India

Shweta Dixit
Clinical Development Services Agency, Translational Health Science and Technology Institute, NCR Biotech Science Cluster, Haryana, India

Fozia Homa
Department of Statistics, Mathematics and Computer Application, Bihar Agricultural University, Sabour, Bhagalpur, Bihar, India

Mukti Khetan
Department of Mathematics, Amity School of Applied Sciences, Amity University, Mumbai, Maharashtra, India

Vijay Kumar
Department of Mathematics, Indian Institute of Technology Bombay, Powai, Mumbai, Maharashtra, India

Pradeep Mishra
College of Agriculture, Rewa, JNKVV, Jabalpur, Madhya Pradesh, India

Ashok Kumar Pathak
Department of Mathematics and Statistics, School of Basic and Applied Sciences, Central University of Punjab, Bathinda, Punjab, India

Shubham Thakur
Department of Statistics, Mathematics and Computer Application, Bihar Agricultural University, Sabour, Bhagalpur, Bihar, India

Prashant Verma
Department of Statistics, Faculty of Science, University of Allahabad, Prayagraj, India

Abbreviations

AUC	area under curve
CDF	cumulative distribution function
CF	characteristic function
CGF	cumulant generating function
CPI	cumulative performance index
FN	false negative
FP	false positive
GSK	GlaxoSmithKline
MGF	moment generating function
MTTF	meantime to failure
PDF	probability density function
PGF	probability generating function
PMF	probability mass function
ROC	receiver operating characteristic
RV	random variable
TN	true negative
TP	true positive

Preface

The book is written in a way that it has the least prerequisites for its readers. So, anybody with a basic graduate-level understanding of statistics would benefit from it. The book's ultimate aim is to provide an understanding of distribution theory and data analysis through statistical software. It will prove helpful to readers in solving problems related to basic statistics, probability models, and simulation using statistical software. It will also explains statistical data analysis by taking multiple real-life situations using the open-source software R version 4.0 and Python 3.0+. A detailed study of the statistical models has been provided with examples related to health, agriculture, insurance, etc. Each chapter will help increase your knowledge, starting from basic statistics to advance statistics.

This book comprises seven chapters. Chapter 1 introduces statistics to the reader, including definition, scope utilization, and its importance in day-to-day life. This chapter further discusses the methods for graphical representations and summary statistics with the help of numerous examples related to real-life situations.

Chapter 2 establishes the concept of random variables (RVs) and their probability distributions, the classification of RVs based on distribution function, and the shapes of the distributions. It also talks about several statistical properties along with the theorems. This chapter profoundly describes the properties with suitable examples.

Chapter 3 presents several discrete probability models and their properties, along with detailed derivations. All the mentioned distributions have been explained with their importance in real-life situations, and simulation methods using statistical software R.

Chapter 4 consists of the concepts of continuous distributions and derivation of their properties. Further in this chapter, the generation of random samples from the mentioned continuous distribution using R software is introduced for readers to achieve practical insights.

Chapter 5 introduces the family of the Weibull distributions with its different cases along with its characteristics such as reliability function, hazard function, mean time to failure (MTTF), etc. Some modifications to the Weibull family have been provided. This chapter also discusses the process of sample generation from the Weibull family of distributions using R-software.

Chapter 6 presents a few life distributions, such as Pareto, generalized Pareto, and Burr distributions, along with their detailed properties. These properties have been discussed with various graphs generated over R-software for better understanding.

Chapter 7 introduces data analysis through the freely available statistical package Python. The chapter is written from the perspective of statistical analysis rather than from the perspective of statistical computing through Python. It will help readers to quick start their data analysis journey through the advanced packages of Python software. The exploratory data analysis, considering the open-source titanic data, explains all aspects of data analysis, right from data cleaning to data interpretation with statistical modeling.

CHAPTER 1

Descriptive Statistics: Principles and Applications

FOZIA HOMA[1], MOHD. ARSHAD[2], VIJAY KUMAR[3], SHUBHAM THAKUR[1], and PRADEEP MISHRA[4]

[1]*Department of Statistics, Mathematics, and Computer Application, Bihar Agricultural University, Sabour, Bhagalpur, Bihar, India*

[2]*Department of Mathematics, Indian Institute of Technology Indore, Simrol, Indore, Madhya Pradesh, India*

[3]*Department of Mathematics, Indian Institute of Technology Bombay, Powai, Mumbai, Maharashtra, India*

[4]*College of Agriculture, Rewa, JNKVV, Jabalpur, Madhya Pradesh, India*

1.1 INTRODUCTION

Statistics have been described or defined differently from time to time by many authors. In ancient times, statistics was only limited to affairs of state, but now the use of statistics has been widened considerably, and thus, with time, the old definitions have been replaced by new definitions which are more detailed and exhaustive. Thus, the most detailed and exhaustive definition has been given by Prof. Horace Secrist as he defined statistics as an *"Aggregate of facts affected to a marked extent by a multiplicity of causes, numerically expressed, enumerated or estimated according to*

Distribution Theory: Principles and Applications. Fozia Homa, Mukti Khetan, Mohd. Arshad, & Pradeep Mishra (Naz) (Eds.)

reasonable standards of accuracy, collected in a systematic manner for a predetermined purpose and placed in relation to each other."

In the case of traditional definitions, Dr. A. L. Bowley has defined statistics as "*Statistics may be called the science of counting.*" This definition does not define the statistics properly as in the case of large numbers, we cannot count them, but we estimate them. In another definition, he said, "*Statistics is the science of measurement of social organism regarded as a whole in all its manifestations.*" The definition is vague as this nature of statistics is not defined properly, and the applications of statistics are limited to the social aspect only, or we can say to humans and its activities. There are many other definitions given by different scientists. Some are listed below:

- "*Statistics is the science of estimates and probabilities.*"—*Boddington*
- "*Statistics may be regarded as a body of methods for making wise decisions in the face of uncertainty.*"—*Wallis and Roberts*
- "*Statistics is defined as the science of collection, presentation, analysis, and interpretation of numerical data.*"—*Croxton and Cowden*
- "*Statistics is both a science and an art.*"—*L. H. C. Tippet*

1.1.1 USES OF STATISTICS

Statistics can be used in every aspect of life in one way or other. Statistics can be used in education, sports, and health, by the government in decision-making, data science, robotics, business, weather forecasting, and much more. Statistics can be used in our day-to-day life to calculate the right time for getting up to go to the office, school, college, or any program. Similarly like, how much money is required for my monthly expenses or to reach my school and office in a month? In education for comparing the marks of students or to check the performance of students. In health like to calculate how many peoples are affected by a particular disease or to calculate average no. of accidents in an area on a yearly basis. In sports for checking his performance over a period of time and compare it with others in a particular game. Statistics are also used in machine learning, deep learning, aerospace, and many other aspects. Nowadays, several advanced techniques of statistics have been developed for big data analytics.

1.1.2 STATISTICS WITH OTHER DISCIPLINES

Statistics has relation almost with every discipline. I can't even name them as the utility of statistics is very wide and it covers every discipline and every sphere of life is affected by statistics in one way or another. Earlier it was only limited to the affairs of the state, but with time, statistics is diversified into various disciplines – physical and natural sciences, social, economics, business, state, and administration, planning, biology, agriculture, astronomy, psychology, and many more. Statistics is mainly applied to quantitative data, but we can also use statistics in the case of qualitative data; for example, attributes such as deaths, defective, new, and old can be transformed into numbers of deaths or number of defectives, etc., as we can quantify them and apply statistics.

1.1.3 ABUSES OF STATISTICS

Statistics are very useful in every aspect of life, but as with good, some bad things also occur in every day-to-day life. Similarly, statistics has some abuses also. One can use the knowledge of statistics for his benefit by manipulating data or sneaky tricks to mislead the naïve. Bad data collection–collecting irrelevant samples, small samples, biased samples, non-random sampling methods, and using a proxy to replace the data that was lost or not available. Poor application of statistical methods–writing wrong formulae, poor mathematics, use of the wrong statistic, misreporting of errors. Loaded questions–by manipulating the words of question to elicit a certain answer from the respondent. Misleading graphs, pictures, and ignoring important facts and features.

1.2 PRESENTATION OF DATA

The first step in statistical analysis is to collect the data and data are generally collected in raw format and which is difficult to understand so after the organization of data the main thing is how we are going to present the data, which is easily understandable to both author and readers and suitable for further analysis and to make suitable conclusions. Data are presented mainly in the following ways:

1. **Textual Presentation:** Data is presented in the form of texts and one can reach conclusions by reading the collected data and suitable when the quantity is not very large. This method requires a long time to reach conclusions as we have to read the whole text, which is quite boring and not suitable for quick decision making.
2. **Tabular Presentation:** It is the orderly and systematic arrangement of data in rows and columns. Main components of the table are:

   i. Table number
   ii. Title
   iii. Headnote
   iv. Column heading
   v. Body
   vi. Stub heading
   vii. Row headings
   viii. Footnote
   ix. Source

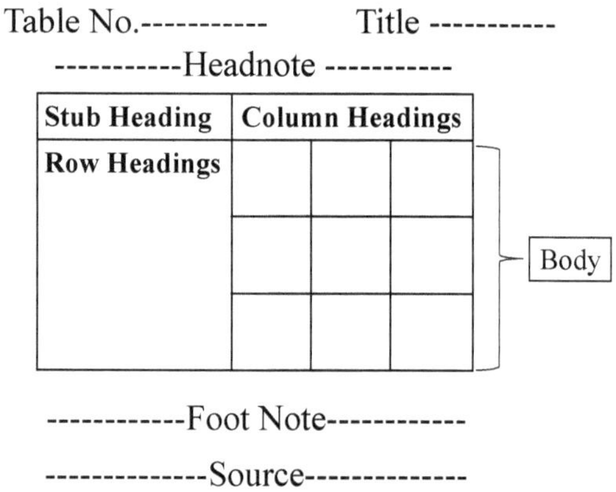

3. **Graphical or Diagrammatic Presentations:** Data shown in tables can be presented in a better way by diagrams and graphs. A diagram is a symbolic or pictorial representation of data and graphs are modifications of diagrams where the mathematical relation between variables can be compared. A graph has two axes horizontal x-axis/ abscissa and vertical y-axis/ordinate. Some standard presentations are:

   i. **Pie Chart:** A pie chart (or a circle chart) is a circle divided into component sectors with areas proportional to the size of the corresponding component (Figure 1.1). Pie charts are widely used in business and media for comparing the different data sets. The followings are the key points for a pie chart:

      a. A pie chart is presenting the data in a circular graph;
      b. Pie charts are the frequently used graphs to represent real-world information;
      c. Pie charts are suitable for discrete data (qualitative data);
      d. The complete pie chart represents the whole data, and their slices represent the parts of the whole data.

➢ **Example 1.1:** The following data represents confirmed cases of COVID-19 of top five States in India till April 22, 2021. The data were collected from www.mygov.in.

State	Total Confirmed Cases	Percentage (approx.)
Maharashtra	4,094,840	50
Kerala	1,322,054	16
Karnataka	1,247,997	15
Uttar Pradesh	9,76,765	12
Chhattisgarh	6,05,568	7
Total	**8,247,224**	**100**

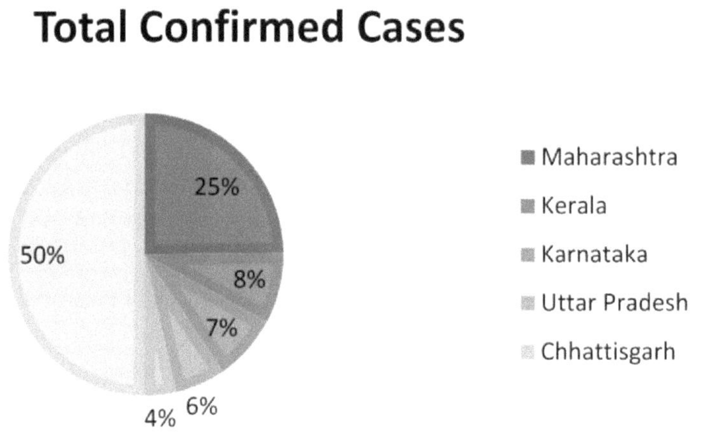

FIGURE 1.1 Pie chart.

ii. **Line Diagram:** It is a chart that shows a line joining several points. It is useful to represent time-series data. Suppose (t,y_t) denote a time series data, y_t represents the value of the variable at time t. We plot each pair of data in a two-dimensional plane and then joined the successive points by line segments. Then the resulting chart is a line diagram (Figure 1.2).

➢ **Example 1.2:** The following data represents COVID-19 cumulative death cases from date (09/04/2021–23/04/2021). The data were picked up from the Ministry of Health and Family Welfare, India.

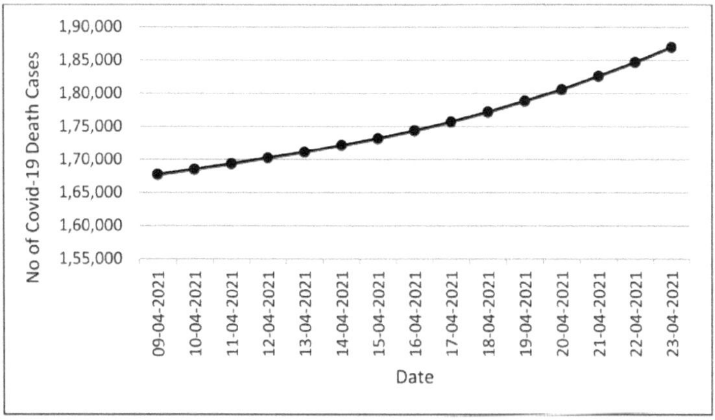

FIGURE 1.2 Line chart.

ii. **Bar Graph:** It displays the distribution of a qualitative (categorical) variable, showing the counts for each category next to each other for easy comparison. Bar graphs should have small spaces between the bars to indicate that these are freestanding bars that could be rearranged into any order (Figures 1.3 and 1.4). The followings are the key points for a bar graph:

   a. A bar graph is a way of demonstrating data using bars in which the length of each bar is proportional to the value they represent;
   b. The bars present a visual display for assessing quantities in different categories;
   c. The bars are lined up along a common base;
   d. Each bar should have equal width;
   e. Bar graphs are useful in data handling.

➢ **Example 1.3:** The following data represents COVID-19 cases of India. The data were taken from https://en.wikipedia.org/wiki/COVID-19_pandemic_in_India.

COVID-19 Dashboard	
Total positive cases	**15,924,732**
New samples tested	1,639,357
Total active cases	2,290,689
Total recovered cases	13,449,371
As of April 21, 2021	

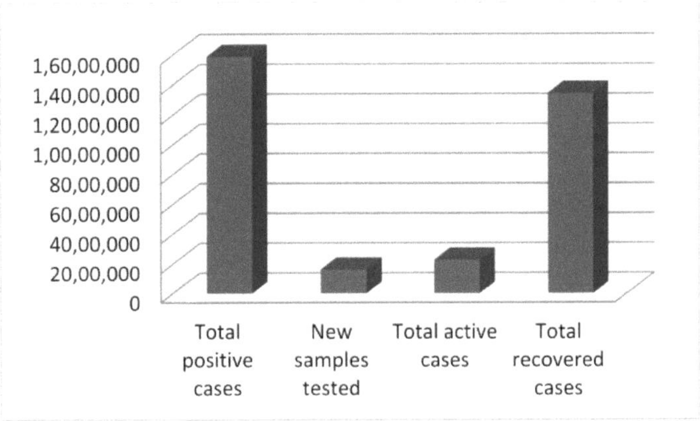

FIGURE 1.3 Simple bar chart.

COVID-19 Pandemic in India by State			
State/Union Territory	**Cases**	**Recoveries**	**Active**
Karnataka	1,247,997	1,037,857	196,255
Kerala	1,322,054	1,160,472	156,554
Uttar Pradesh	976,765	706,414	259,810
West Bengal	700,904	621,340	68,798
Maharashtra	4,094,840	3,330,747	701,614
As of 23 April 2021			

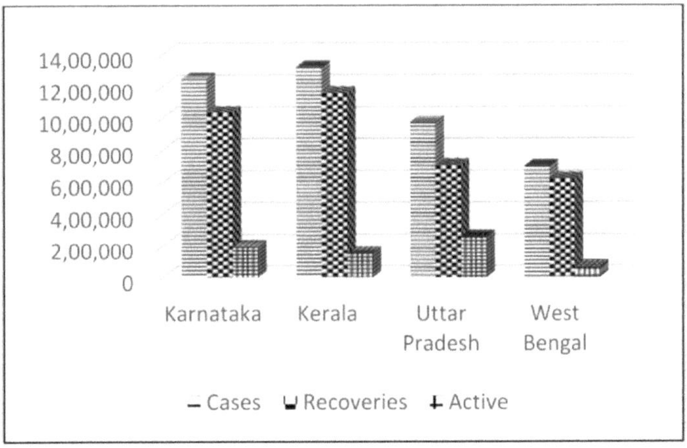

FIGURE 1.4 Multiple bar chart.

The data were taken from: https://en.wikipedia.org/wiki/Template: COVID19_pandemic_data.

iv. **Histogram:** It is the graphical presentation of data where data is grouped into continuous number ranges and each range corresponds to a vertical bar. Firstly, we will discuss the groping of data by using the concept of frequency distributions (Figure 1.5).

- **Frequency Distributions:** It is the distribution of the total number of observations into different classes, which shows all the distinct observations of our interest and the number of times they occur (frequency of occurrence). It is generally of two types:
    - **Discrete Frequency Distribution:** Suppose we have an individual series with a finite number of observations. When the number of observations is large and the range of data is small, we convert the individual series into discrete frequency distribution.
    - **Continuous Frequency Distribution:** When the number of the observations is large and range of data is also large, we convert it into class limits (lowest and highest values that can be included in the class) with some class interval (difference between upper and lower limit) such type of frequency distribution is called continuous frequency distributions. Numbers of classes included can be calculated by $K = 1 + 3.322\log N$ (base 10) where N is the total frequency (number of observations).

Graphical representation of frequency distribution can be in the form of histogram and frequency polygon. In the case of continuous frequency distribution, we take class intervals along the x-axis and frequency on the y-axis. In each class interval, we draw an erect rectangle height equal to the frequency of that class. Thus, this diagram of continuous rectangles is known as a histogram. If the frequency distribution is not continuous, it is not an appropriate diagram to represent the data. However, some alternative diagrams can be drawn in such cases as bar diagrams. We cannot construct a histogram for distribution with open-end classes.

➢ **Example 1.4:** The following data represents glucose blood level (mg/100 ml) after a 12 hour fast for a random sample of 70 women. (The data were picked up from American J. Clin. Nutr. Vol. 19, 345–351).

Glucose	Frequency
45–55	3
55–65	6
65–75	18
75–85	29
85–95	9
95–105	4
105–115	1

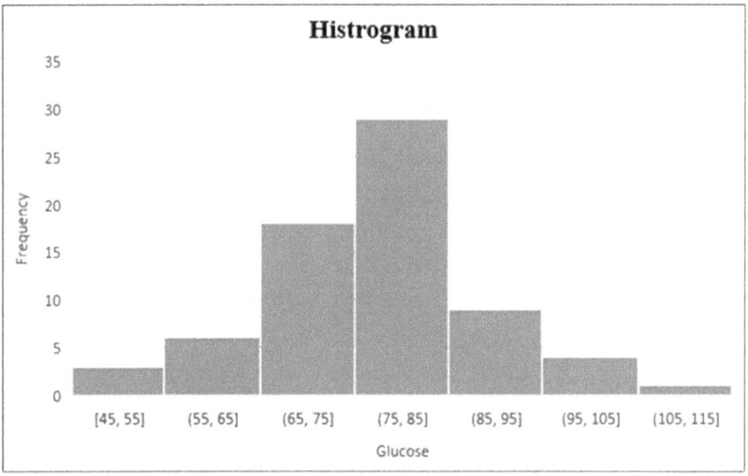

FIGURE 1.5 Histogram.

v. **Frequency Polygon:** In the case of ungrouped data the frequency polygon can be obtained by plotting the variate values on the x-axis and their corresponding frequency on the y-axis and by joining the points by straight lines. In the case of grouped data with equal class intervals on the x-axis, we take mid values of the class, and on the y-axis take frequency and join the midpoints using straight lines (Figure 1.6).

- **Example 1.5:** The following frequency polygon is prepared by using the data given in Table 1.1 (Glucose blood level (mg/100 ml) after a 12 hour fast).

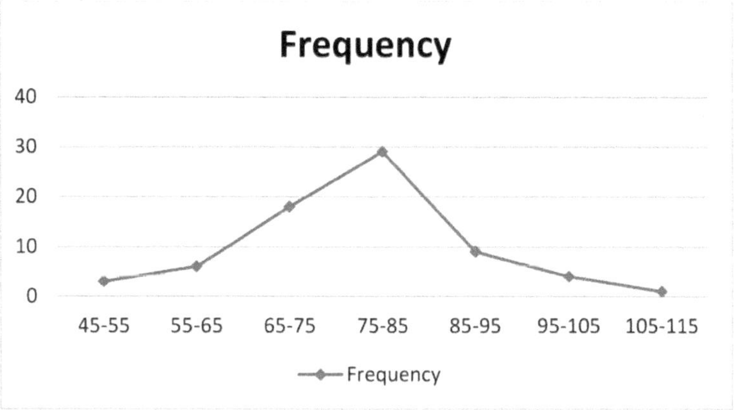

FIGURE 1.6 Frequency polygon.

vi. **Cumulative Frequency Curve or Ogive:** Firstly, we will discuss the preparation of cumulative frequency distribution. Cumulative frequency is the running total of frequencies up to a certain point (or class) and the distribution of cumulative frequency is known as cumulative frequency distribution. It can be graphically represented by a cumulative frequency curve or ogive. Cumulative frequency is of two types:

   a. Less than cumulative frequency: It is the sum of the frequency of that class and the frequencies of all preceding classes; and
   b. More than cumulative frequency: It is the sum of the frequency of that class and all the frequencies of classes below it.

Cumulative frequency curve (or ogive) is obtained by plotting upper or lower limits of class interval on the x-axis and its corresponding frequency on the y-axis. There are two methods of constructing ogive:

- **Less Than Type:** If the group is not continuous, we first convert it into a continuous group by subtracting d/2 from the lower limit of class and adding d/2 to the upper limit of class where d is the jump from the upper limit of the preceding class to the lower limit of next class. Then by plotting the upper limit of a class on the x-axis against the corresponding less than cumulative frequency on the y-axis, we get some points and thus by free-handedly joining these points we get that less than type ogive (Figure 1.7).

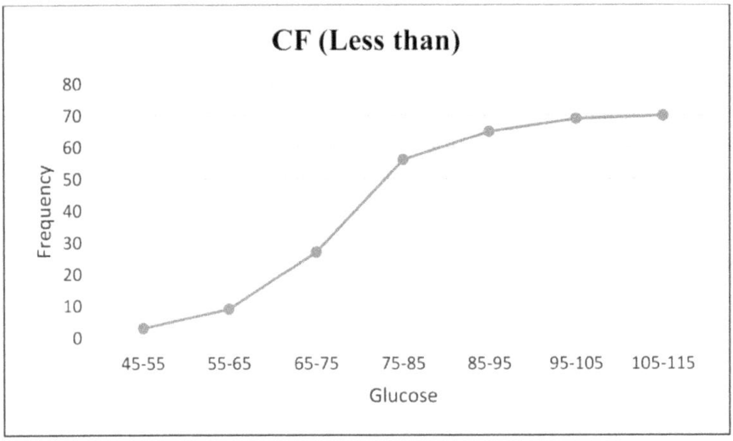

FIGURE 1.7 Less than type ogive.

- **More Than Type:** Here we take lower limits of the class on the x-axis and corresponding more than cumulative frequency on the y-axis, and by joining the points we get the more than type ogive (Figure 1.8).

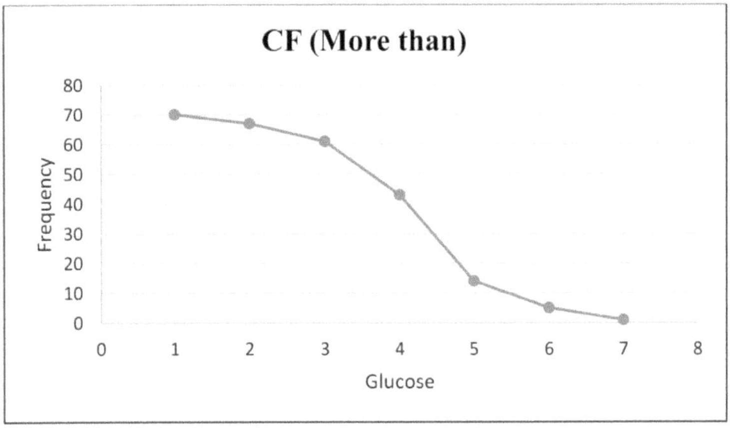

FIGURE 1.8 More than type ogive.

From ogive (less than and more than types) we can find the value of the median. If we draw a perpendicular line at the point of intersection of less than type and more than type on the x-axis where the perpendicular line meet is the value of the median (Figure 1.9).

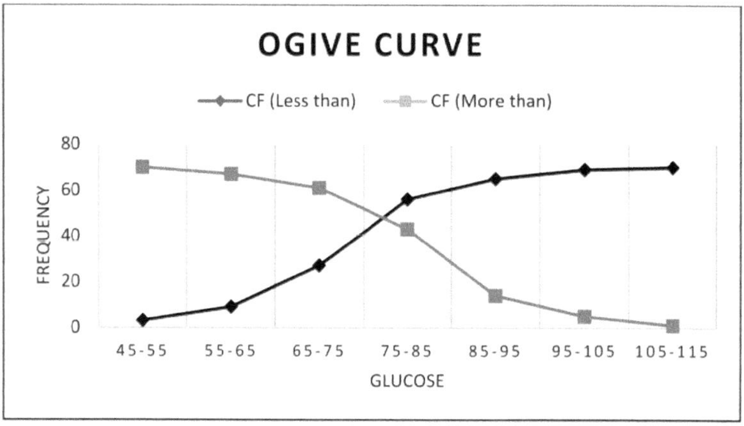

FIGURE 1.9 Ogive.

➢ **Example 1.6:** The following ogives (less than and more than type) are prepared by using the data given in Table 1.1.

TABLE 1.1 Glucose Blood Level (mg/100 ml) After a 12 Hour Fast

Glucose	CF (Less than)	CF (More than)
45–55	3	70
55–65	9	67
65–75	27	61
75–85	56	43
85–95	65	14
95–105	69	5
105–115	70	1

vii. **Scatter Plot:** It is a dots plot to represent the values of two variables. It is used to investigate the relation between two variables. In this plot, the first variable assigned to the horizontal axis, and another variable assigned to the vertical axis. It helps to determine whether the relationship between two variables is linear. It also provides a way to detect outliers in the data set (Figures 1.10 and 1.11).

- **Example 1.7:** Auto insurance in Sweden data. In the following data: X = Number of claims;

  Y = Total payment for all the claims in thousands of Swedish Kronor for geographical zones in Sweden.

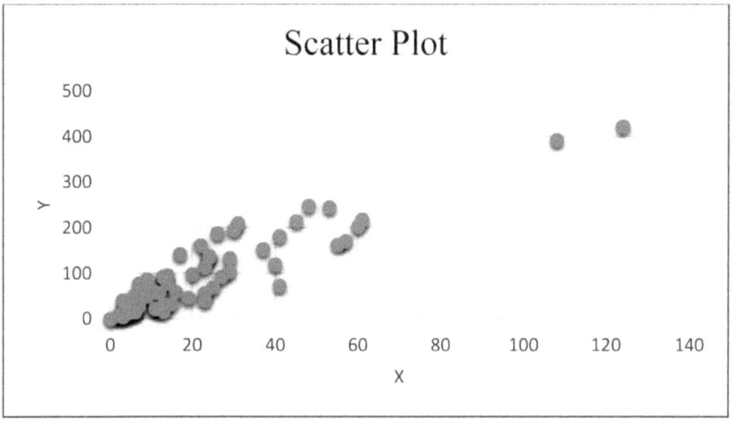

FIGURE 1.10 Scatter plot.

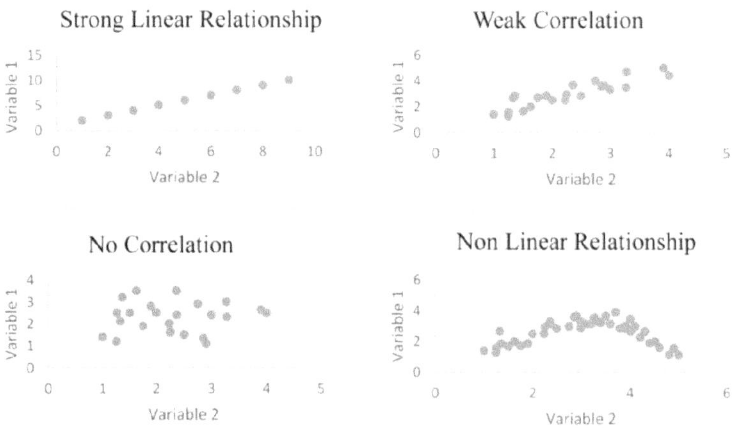

FIGURE 1.11 Correlation analysis through scatter plot.

Reference: Swedish Committee on Analysis of Risk Premium in Motor Insurance.

Source: https://college.cengage.com.

1.3 DESCRIPTIVE ANALYSIS

Descriptive analysis is a classical approach to data analysis. It is a very useful statistical tool to provide good conclusions for a real-life problem based on data information. It includes different measures of central tendency, dispersion, skewness, kurtosis, etc. These measures help us to analyze the data mathematically. Here, we will discuss these measures in brief.

1.3.1 MEASURE OF CENTRAL TENDENCY

An average is a single value within the range of the data that is representative of the data. Since an average is somewhere within the range of the data, it is also called a Measure of Central Tendency. In statistics, central tendency represents the center or location of the distribution of the data. Central tendency allows researchers to summarize a large set of data into a single number. It is possible to compare two (or more) sets of data by simply comparing the measure of central tendency for different data sets. A good measure of central tendency (or average) should be rigorously defined, depends on all the observations, less influenced by extreme observations, capable of interpretation, easy to use for further algebraic treatments, etc. Table 1.2 provides a different measure of central tendency.

TABLE 1.2 Measure of Central Tendency

Mathematical Average	Positional Average	Commercial Average
Arithmetic mean	Median	Moving average
Geometric mean	Mode	Progressive average
Harmonic mean	Quartile, decile, percentile	Composite average

1. **Arithmetic Mean:** Mean of a set of observations is the sum of observations divided by the total number of observations. Let $x_1, x_2, \ldots x_n$ be the observations in an experiment, then the arithmetic mean (or mean) of these observations is given by:

$$\bar{X} = \frac{\sum_{i=1}^{n} x_i}{n}$$

In case of frequency distribution, the arithmetic mean of the data is defined as:

$$\overline{X} = \frac{\sum_{i=1}^{n} f_i x_i}{\sum_{i=1}^{n} f_i}$$

where; f_i denotes the frequency of the observation x_i, $i = 1, 2, \ldots, n$. The detailed calculation is given in the Table 1.3.

➢ **Example 1.7:** Calculate the arithmetic mean for data of glucose blood level (mg/100 ml) given in Tables 1.1 and 1.3.

TABLE 1.3 Calculation of Arithmetic Mean

Glucose	Frequency (f)	Mid-Value (x)	fx
45–55	3	50	150
55–65	6	60	360
65–75	18	70	1,260
75–85	29	80	2,320
85–95	9	90	810
95–105	4	100	400
105–115	1	110	110
Total	**70**	–	**5,410**

$$\overline{X} = \frac{\sum_{i=1}^{n} f_i x_i}{\sum_{i=1}^{n} f_i} = \frac{5410}{70} = 77.29.$$

2. **Geometric Mean:** It is the n th root of the product of n values of a set of observations. It is appropriate for averaging the ratios of changes, for an average of proportions, etc. Let $x_1, x_2, \ldots x_n$ be the observations in an experiment, then the geometric mean of these observations is given by:

$$GM = (x_1 . x_2 . \ldots . x_n)^{\frac{1}{n}}$$

If we take log both side of the above formula and simplify it, then the formula for the geometric mean is given by:

$$GM = Antilog\left[\frac{1}{n}\left(log(x_1) + log(x_2) + \cdots + log(x_n)\right)\right]$$

In case of frequency distribution, the geometric mean of the data is defined as:

$$GM = Antilog\left[\frac{1}{\sum_{i=1}^{n} f_i}\left(f_1 * log(x_1) + f_2 * log(x_2) + \cdots + f_n * log(x_n)\right)\right]$$

where; f_i denotes the frequency of the observation x_i, $i = 1, 2, \ldots, n$.

➢ **Example 1.8:** Consider a stock that grows by 15% in the first year, declines by 10% in the second year, and then grows by 20% in the third year. The geometric mean of the growth rate is calculated as follows:

$$GM = \left((1+0.15)*(1-0.1)*(1+0.2)\right)^{\frac{1}{3}}$$
$$= 1.075$$
$$= 1 + 90.075$$

or 7.5% growth rate annually.

3. **Harmonic Mean:** It is the inverse of the arithmetic mean of the reciprocals of the observations. It is appropriate for the rate of change per unit time such as speed, number of items produced per day, etc. Let $x_1, x_2, \ldots x_n$ be the given observations, then the harmonic mean of these observations is given by:

$$HM = \frac{n}{\frac{1}{x_1} + \frac{1}{x_2} + \cdots + \frac{1}{x_n}}$$

In case of frequency distribution, the geometric mean of the data is defined as:

$$HM = \frac{f_1 + f_2 + \cdots + f_n}{\frac{f_1}{x_1} + \frac{f_2}{x_2} + \cdots + \frac{f_n}{x_n}}$$

where; f_i denotes the frequency of the observation x_i, $i = 1, 2, \ldots, n$.

➢ **Example 1.9:** If a family purchases the milk per liter during October, November, and December is Rs. 50, Rs. 48, and Rs. 45 per liter, then the average price of the milk in these three months will be:

$$HM = \frac{3}{\frac{1}{50} + \frac{1}{48} + \frac{1}{45}} = 47.58$$

4. **Median:** It is that value that divides the whole data into two equal parts. It has half of the values less than the median and half of the values greater than the median. It is a positional average and is not affected by the extremely large or small values in the data. It can be computed for ratio, interval, and ordinal data but not appropriate for nominal data. If the distribution is skewed, the median is a better measure as compared to the arithmetic mean. But it is not capable of further algebraic treatment, e.g., if the medians of two or more data sets are given, then the combined median cannot be calculated, whereas in such cases, the combined mean can be calculated.

   Let x_1, x_2,…x_n be the given observations, then the median of these observations can be calculated as follows:

   - First, arrange the observations in increasing (or decreasing) order of magnitude, i.e., let $x_{(1)} \le x_{(2)} \le \ldots \le x_{(n)}$ be the ordered values of $x_1, x_2, \ldots x_n$.
   - If the sample size n is an odd number, then $\left(\frac{n+1}{2}\right)th$ observation in the order data is the median.
   - If the sample size n is an even number, then the median is the average of $\left(\frac{n}{2}\right)th$ observation and $\left(\frac{n}{2}+1\right)th$ observation in the order data.

➢ **Example 1.10:** Find the median in the following data:

a. 4, 7, 2, 8, 9, 10, 8.
   First, arrange data in increasing order, i.e., 2, 4, 7, 8, 8, 9,10. The sample size n = 7 is an odd number, i.e., $\left(\frac{7+1}{2}\right) = 4th$ observation 8 is the median.

b. 28, 30, 10, 12, 3, 6, 35, 38, 17, 40.
   First, arrange data in increasing order, i.e., 3, 6, 10, 12, 17, 28, 30, 35, 38, 40.The sample size n =10 is an even number. Here, $\left(\frac{10}{2}\right) = 5th$ observation is 17, and $\left(\frac{10}{2}+1\right) = 6th$ observation is 28. Thus, the median of the given data is $\frac{17+28}{2} = 22.5$.

5. **Mode:** It is that value that is occurring more frequently or having a maximum frequency. It is appropriate for any measurement level, i.e., nominal, ordinal, interval, and ratio. It is not necessarily to be unique.

A distribution with a single mode is said to be *unimodal.* If two or more values occur the same number of times, then there will be two or more modes. A distribution with more than one mode is said to be *bimodal, trimodal*, etc., or in general, *multi-modal.* If no value is repeated, no mode is defined (Figure 1.12).

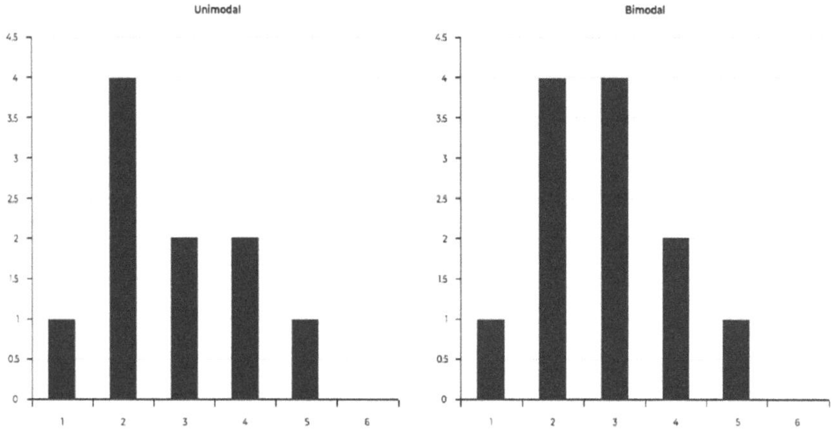

FIGURE 1.12 Unimodal and bimodal distributions.

- **Example 1.11:** Gestational ages (weeks) of 14 newborns: 32, 36, **39, 39,** 42, 35, 40, 34, **39,** 41, 33, 38, 37, 39. There are four observations with a gestation age of 39 weeks, the rest of the values are occurring only once. Thus, the mode of the given data is 39 weeks.

6. **Quartiles:** The three points which divide the data into four equal parts are called quartiles. The first, second, and third points are known as the first, second, and third quartiles, respectively. It is a generalization of the idea of the median. The first quartile Q_1 is the value which exceeds 25% of the observations and is exceeded by 75% of the observations. The second quartile Q_2 coincides with the median. The third quartile Q_3 is the point which has 75% observations before it and 25% observations after it. The 5-number summary of data reports its median, quartiles, and extremes observations, i.e., the maximum observation and the minimum observation.
7. **Deciles:** The nine points which divide the data into 10 equal parts are called deciles. For example, the seventh decile D_7 has 70% observations

before it and 30% observations after it. The fifth decile D_5 coincides with the median.

8. **Percentiles:** The 99 points which divide the data into 100 equal parts are called percentiles. For example, the 47th percentile P_{47} has 47% observations before it and 53% observations after it. The 50th percentile P_{50} coincides with the median. Percentiles indicate the percentage of scores that fall below a particular value. They tell us where a score stands relative to other scores.

1.3.2 MEASURE OF DISPERSION

It measures the scatteredness and variability of the mass of figures in given data from its average. There are mainly two types of dispersions:

1. **Absolute Measure of Dispersion:**
   i. **Range:** It is the difference between the maximum and minimum values in the given set of data. Let $x_1, x_2, \ldots x_n$ be the given data, and let $x_{(1)} \leq x_{(2)} \leq \ldots \leq x_{(n)}$ be the ordered values of $x_1, x_2, \ldots x_n$. Then, the range is defined as:
   $Range = x_{(n)} - x_{(1)}$
   ii. **Quartile Deviation:** It is half of the interquartile range, i.e., $Q_D = \frac{Q_3 - Q_1}{2}$, where $Q_3 - Q_1$ is the interquartile range, and Q_1 and Q_3 are first and third quartile.
   iii. **Mean Deviation:** It is the arithmetic mean of the absolute value of the deviation of observation from its average value (mean, median, or mode). Let $x_1, x_2, \ldots x_n$ be the given observations, then the mean deviation about an average A (mean, median, or mode) is given by:

   $$MD(A) = \frac{1}{n}\sum_{i=1}^{n}|x_i - A|$$

   vi. **Variance and Standard Deviation:** The variance measures the spread between the observations in the data set. It is defined as the average of the square of deviation of observations from its mean value.). Let $x_1, x_2, \ldots x_n$ be the given observations, then the variance is denoted by σ^2 and is given by:

$$\text{Variance} = \frac{1}{n}\sum_{i=1}^{n}(x_i - \bar{x})^2$$

where; $\bar{x}$ denotes the mean of the data set. The positive square root of the variance is called the standard deviation. It is denoted by σ. Note that the above formula for variance is an estimator of population variance and has some undesirable properties. To overcome these, we may define the sample variance as:

$$\text{Corrected Variance} = \frac{1}{n-1}\sum_{i=1}^{n}(x_i - \bar{x})^2$$

This corrected variance is a useful measure of variability in the analysis of real-life data.

2. **Relative Measure of Dispersion:** When two or more than two series to be compared which differ in their units, the relative dispersion is calculated by making the series unit free. The coefficient of dispersion is the measure of relative dispersion.

- Coefficient of dispersion based on range:

$$COD\ (range) = \frac{L-S}{L+S}$$

where; L is the maximum observation and S is the minimum observation.

- Coefficient of dispersion based on quartile deviation:

$$COD\ (Q_D) = \frac{Q_3 - Q_1}{Q_3 + Q_1}$$

where; Q_1 is the first quartile; and Q_3 is the third quartile.

- Coefficient of dispersion based on mean deviation:

$$COD(MD) = \frac{Mean\ Deviation}{Average\ value(mean, mode\ or\ median)}$$

- Coefficient of dispersion based on standard deviation:

$$CD\ (\sigma) = \frac{\sigma}{\overline{\text{X}}}$$

where; σ is the standard deviation; and $\overline{X}$ is the mean.

- Coefficient of variation: Around 100 times the coefficient of dispersion based on the standard deviation is known as the coefficient of variation. It measures the percent variation.

$$C.V = \frac{\sigma}{\overline{X}} * 100$$

The series with smaller C.V is more consistent and has less variation as compared to series with higher C.V and have more variation.

1.3.3 MEASURE OF SKEWNESS

It is a measure of lack of symmetry in the shape of a distribution. If the mean, median, and mode are not the same, then distribution is said to be skewed and quartiles are not equidistant from the median. Skewness is of two types (Figure 1.13):

i. **Positively Skewed:** A series is said to be positively skewed if mean > median > mode and more frequencies lie towards the right-hand side of the distribution.

ii. **Negatively Skewed:** A series is said to be negatively skewed if mode > median > mean and more frequencies lie toward the left-hand side of the distribution.

We may also identify the skewness of the distribution of the data by using the concept of third sample moment. Let $x_1, x_2, \ldots x_n$ be the given observations, then third sample moment about the mean is defined as:

$$\mu_3 = \frac{1}{n}\sum_{i=1}^{n}(x_i - \bar{x})^3$$

where; $\bar{x}$ denotes the mean of the data set. If $\mu_3 = 0$, the distribution of data is symmetric, otherwise it is skewed. If $\mu_3 > 0$, the distribution is positively skewed, and if $\mu_3 > 0$, the distribution is negatively skewed.

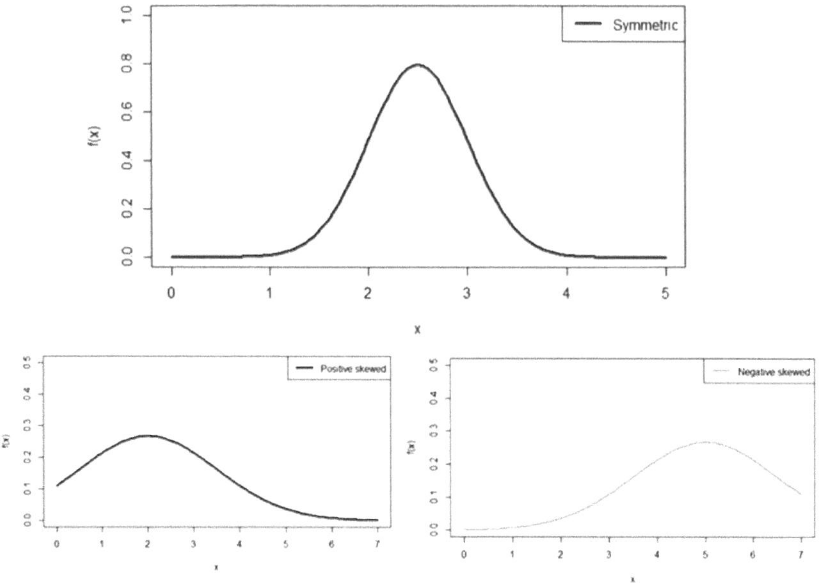

FIGURE 1.13 Symmetric, positive skewed, and negative skewed.

1.3.4 *MEASURE OF KURTOSIS*

It measures the flatness or peak of the distribution and tells whether the distribution is flatter or peaked. It also provides a way to identify the data that are heavy-tailed or light-tailed relative to a normal distribution. The curve of the normal distribution is called mesokurtic. The data distribution having greater kurtosis than the normal is called leptokurtic. Similarly, the data distribution having smaller kurtosis than the normal is called platykurtic. Let $x_1, x_2, \ldots x_n$ be the given observations, then second sample moment about the mean (variance) and fourth sample moment about the mean are, respectively, defined as:

$$\mu_2 = \frac{1}{n}\sum_{i=1}^{n}(x_i - \bar{x})^2$$

and

$$\mu_4 = \frac{1}{n}\sum_{i=1}^{n}(x_i - \bar{x})^4$$

where; $\bar{x}$ denotes the mean of the data set. Now, the coefficient of kurtosis is defined by (Figure 1.14):

$$\beta = \frac{\mu_4}{\mu_2^2}$$

It is a number without a unit of measurement. If $\beta = 3$, the distribution of data is mesokurtic (normal). If $\beta < 3$, the distribution is leptokurtic, and if $\beta < 3$, the distribution is platykurtic.

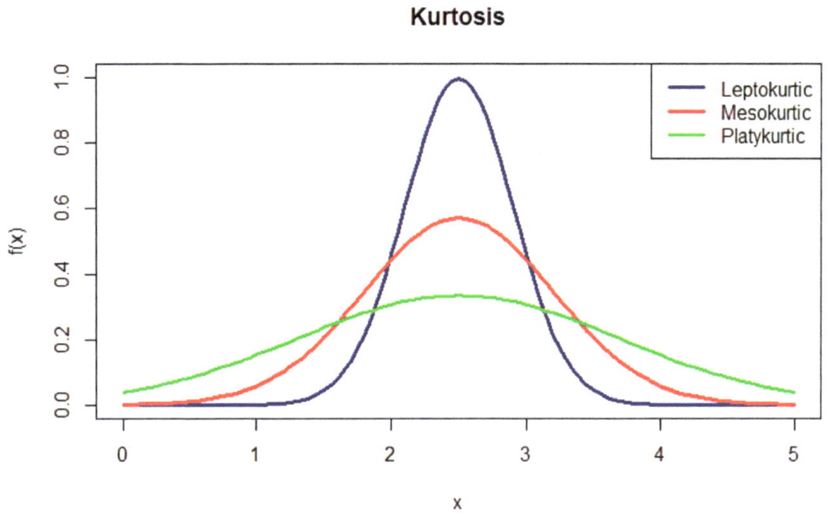

FIGURE 1.14 Kurtosis of the distribution.

KEYWORDS

- **descriptive analysis**
- **frequency distributions**
- **histogram**
- **statistical analysis**
- **tabular presentation**

REFERENCES

Gun, A. M., Gupta, M. K., & Dasgupta, B., (2013). *Fundamentals of Statistics* (Vol. 2). The World Press Pvt. Ltd.: Kolkata.

Gupta, S. C., & Kapoor, V. K., (1997). *Fundamentals of Mathematical Statistics*. Sultan Chand and Sons: New Delhi.

Heumann, C., Schomaker, M., & Shalabh, (2016). *Introduction to Statistics and Data Analysis: With Exercises, Solutions and Applications in R*. Springer.

Miller, I., Miller, M., Freund, J. E., & Miller, I., (2004). *John E. Freund's Mathematical Statistics with Applications*. Prentice Hall: Upper Saddle River, NJ.

Ross, S., (2017). *Introductory Statistics* (4th edn.). Academic Press.

CHAPTER 2

Random Variables and Their Properties

VIJAY KUMAR[1], MUKTI KHETAN[2], MOHD. ARSHAD[3], and SHWETA DIXIT[4]

[1]*Department of Mathematics, Indian Institute of Technology Bombay, Powai, Mumbai, Maharashtra, India*

[2]*Department of Mathematics, Amity School of Applied Sciences, Amity University Mumbai, Maharashtra, India*

[3]*Department of Mathematics, Indian Institute of Technology Indore, Simrol, Indore, Madhya Pradesh, India*

[4]*Clinical Development Services Agency, Translational Health Science and Technology Institute, NCR Biotech Science Cluster, Haryana, India*

2.1 INTRODUCTION

In our daily life, we often face experiments having outcomes with uncertainty. These experiments are called random experiments. The set of all possible outcomes in a random experiment is called the sample space. We use the symbol S for sample space throughout this chapter. Each element (or member) of S has some probability of occurrence. For example, if we appear in a competitive examination, the outcome may be "pass" or "fail" in that examination. Both outcomes (pass or fail) have some probability of occurrence. Let λ_1 and λ_2 denote the outcomes pass and fail, respectively.

Distribution Theory: Principles and Applications. Fozia Homa, Mukti Khetan, Mohd. Arshad, & Pradeep Mishra (Naz) (Eds.)

Then, the sample space is $S = \{\lambda_1, \lambda_2\}$. The subsets of S are called the events. In this example, the class of events is $C = \{\phi, \{\lambda_1\}, \{\lambda_2\}, S\}$. If the probability of an event $\{\lambda_1\}$ (i.e., the result is pass) is 0.65, then the probability of an event $\{\lambda_2\}$ is 0.35, (since $\{\lambda_2\}$ is a complementary event of $\{\lambda_1\}$). Sometimes, S is large and all subsets of S are not important to the experimenter. In such cases, the experimenter collects only the important subsets of S. This class of collected subsets may be extended further to a class $\mathcal{F}$ that contains the subsets of S and satisfying the following properties:

- $S \in \mathcal{F}$;
- $K \in \mathcal{F} \Rightarrow K^c = S - K \in \mathcal{F}$;
- $K_i \in \mathcal{F}, i = 1, 2, \ldots \Rightarrow \bigcup_{i=1}^{\infty} K_i \in \mathcal{F}$.

Then, the class $\mathcal{F}$ is called a sigma-field (or σ-field) of subsets of S. The smallest sigma-field is $\mathcal{F}_1 = \{\phi, S\}$, and the largest sigma-field is the power set of S. Note that the probability measure P is a set function defined on the sigma-field of the subsets of S, i.e., $P: \mathcal{F} \to [0, 1]$ is a set function such that:

- $P(K) \geq 0, \forall K \in \mathcal{F}$;
- $P\left(\bigcup_{i=1}^{\infty} K_i\right) = \sum_{i=1}^{\infty} P(K_i)$, for any sequence of disjoint events $\{K_1, K_2 \ldots\}$ in $\mathcal{F}$;
- $P(S) = 1$.

The above definition is called the axiomatic (or modern) definition of probability. It was given by Kolmogorov in his fundamental work on probability theory. His book "Foundations of the Calculus of Probabilities," published in German in 1933. A Russian version of Kolmogorov's work appeared in 1936, and the first English version was published in 1950 as "Foundations of the Theory of Probability." Kolmogorov's approach is more practical and useful than other earlier formulations of probability theory. It is well known that the probability theory had its origin in gambling and the game of chance. Although, the first mathematical formulation of the probability theory was given by Laplace in 1812. Laplace developed the classical theory of probability, which applied to events that can occur only in a finite number of ways. According to the classical definition, the computation of the probability of events was reduced to combinatorial counting problems. For the more detailed theory of probability and its applications, the readers can see the books by Robert B. Ash; Robert B.

Ash and C. D. Dade. Now, in the next section, we will discuss the random variables (RVs).

2.2 RANDOM VARIABLES (RVS)

Most of the time, the sample space S in a random experiment contains non-numeric outcomes and further analysis of these outcomes is difficult. To overcome this difficulty, we may transform the given S to a subset of the set of real numbers through a function. This idea motivates to define a random variable (RV), which is a function on S that satisfies certain conditions.

Recall that $\mathcal{F}$ denotes the sigma-field of subsets of S in a random experiment, and P denotes the probability measure defined on $\mathcal{F}$. The triplet (S, $\mathcal{F}$, P) is called the probability space.

- **Definition 2.1:** Let (S, $\mathcal{F}$, P) be a probability space. The function $X: S \rightarrow \mathbb{R}$ is said to be a RV if:

$$X^{-1}(-\infty, x] = \{\omega \in S : X(\omega) \leq x\} \in \mathcal{F}, \forall x \in \mathbb{R},$$

where; $\mathbb{R}$ is the set of real numbers.

- **Example 2.1:** A student appears in an examination with five grades, A, B, C, D and F. Each grade represents some points for calculating the cumulative performance index (CPI) of the student for his result. Table 2.1 represents the grading points of the different grades.

TABLE 2.1 Grading Scheme

Grade	**A**	**B**	**C**	**D**	**F**
Points	10	8	6	4	0

Here the sample space $S = \{A, B, C, D, F\}$, and the sigma-field $\mathcal{F} = \mathcal{P}(S)$: the power set of S. The RV $X: S \rightarrow \mathbb{R}$ is given by:

$$X(A) = 10, X(B) = 8, X(C) = 6, X(D) = 4, X(F) = 0$$

This RV X represents the points of the grade awarded by the institute.

Now, we will define a probability function associated with the RV X. Recall that X is a real-valued function and its range set S_X (say) is a subset of $\mathbb{R}$. Since $(S, \mathcal{F}, P)$ is a probability space and X is a RV defined on S, it follows that there exist a class $\mathcal{B}$ (called the Borel-field) of subsets of $\mathbb{R}$ and a probability function P_X (called the induced probability measure) induced by the RV X. The induced probability function P_X is defined as:

$$P_X(B) = P(X^{-1}(B)),\ B \in \mathcal{B}$$

If we take $B = (-\infty, x]$, then the induced probability function is:

$$P_X(\{X \leq x\}) = P_X(-\infty, x] = P(X^{-1}(-\infty, x]),\ \forall\ x \in \mathbb{R}$$

Clearly, $P_X(\{X \leq x\})$ is defined for each $x \in \mathbb{R}$ if P is a probability measure defined on sigma-field. It has many desirable properties and useful to study the various properties of the associated RV X. Now, we define the cumulative distribution function (CDF) of the RV X.

- **Definition 2.2:** Let X be a given RV with induced probability function. Then, the function F: $\mathbb{R} \rightarrow \mathbb{R}$ defined by:

$$F(x) = P_X(\{X \leq x\}),\ \forall\ x \in \mathbb{R}$$

is called the CDF (or distribution function) of RV X.

- **Example 2.2:** (Continued Example 2.1). If we use the past information regarding the result of this examination and calculate the probabilities of the different grades secured by the students. Suppose that Table 2.2 provides the probability distribution.

TABLE 2.2 Probability Distribution of Grades

Grade	**A**	**B**	**C**	**D**	**F**
Probability P(ω)	0.05	0.4	0.35	0.1	0.1

First, we will calculate the induced probability function P_X of the RV X, defined in Example 2.1. We can write P_X in two ways:

i. **Tabular Form:** Table 2.3 provides the induced probability function P_X of the RV X.

TABLE 2.3 Probability Function of the Associated Random Variable

***x* (Value of *X*)**	0	4	6	8	10
Probability $P_X(x)$	0.1	0.1	0.35	0.4	0.05

ii. Functional Form: The following function represents the induced probability function P_X of the RV X.

$$P_X(x) = P_X(X = x) = \begin{cases} 0.1, x = 0, \\ 0.1, x = 4, \\ 0.35, x = 6, \\ 0.4, x = 8, \\ 0.05, x = 10, \\ 0, \text{otherwise}. \end{cases}$$

Now, we will compute the CDF:

$$F_X(x) = P_X(X \le x) = \begin{cases} 0, x < 0, \\ 0.1, 0 \le x < 4, \\ 0.2, 4 \le x < 6, \\ 0.55, 6 \le x < 8, \\ 0.95, 8 \le x < 10, \\ 1, x \ge 10. \end{cases}$$

Note that the CDF $F_X(x)$ of a RV has several nice properties, which help us to analyze the data in real-life experiments. A few are given below:

- F_X is non-decreasing;
- F_X is right continuous at every point in $\mathbb{R}$;
- $\lim_{x\to-\infty} F_X(x) = 0$ and $\lim_{x\to\infty} F_X(x) = 1$.

➢ **Remark:** Let a and b are real constants. The following probability can be calculated, by using the CDF $F_X(x)$, as follows:

i. $P(X < a) = F_X(a^-)$
ii. $P(a \le X \le b) = F_X(b) - F_X(a^-)$
iii. $P(a < X \le b) = F_X(b) - F_X(a)$
iv. $P(a \le X < b) = F_X(b^-) - F_X(a^-)$
v. $P(a < X < b) = F_X(b^-) - F_X(a)$
vi. $P(X = a) = F_X(a) - F_X(a^-)$
vii. $P(X \ge a) = 1 - F_X(a^-)$
viii. $P(X > a) = 1 - F_X(a)$.

➢ **Example 2.3:** If the CDF of a RV X is given by:

$$F_X(x) = \begin{cases} 0, x < -1, \\ \dfrac{x+1}{4}, -1 \le x < 0, \\ \dfrac{x+3}{4}, 0 \le x < 1, \\ 1, x \ge 1. \end{cases}$$

Find the following probabilities:

i. $P\left(-\frac{1}{2} < X \le \frac{1}{2}\right)$

ii. $P\left(-\frac{1}{4} \le X \le \frac{3}{4}\right)$

iii. $P\left(0 < X < \frac{1}{2}\right)$

iv. $P\left(-\frac{1}{2} \le X < 1\right)$

v. $P(X = 0)$

➢ **Solution:** Using the CDF, the following probabilities are as follows:

$$P\left(-\frac{1}{2} < X \le \frac{1}{2}\right) = F_X\left(\frac{1}{2}\right) - F_X\left(-\frac{1}{2}\right) = \frac{\frac{1}{2}+3}{4} - \frac{-\frac{1}{2}+1}{4} = \frac{7}{8} - \frac{1}{8} = \frac{3}{4}$$

$$P\left(-\frac{1}{4} \le X \le \frac{3}{4}\right) = F_X\left(\frac{3}{4}\right) - F_X\left(-\frac{1^-}{4}\right) = \frac{\frac{3}{4}+3}{4} - \frac{-\frac{1}{4}+1}{4} = \frac{15}{16} - \frac{3}{16} = \frac{3}{4}$$

$$P\left(0 < X < \frac{1}{2}\right) = F_X\left(\frac{1^-}{2}\right) - F_X(0) = \frac{\frac{1}{2}+3}{4} - \frac{3}{4} = \frac{7}{8} - \frac{3}{4} = \frac{1}{8}$$

$$P\left(-\frac{1}{2} \le X < 1\right) = F_X\left(1^-\right) - F_X\left(-\frac{1^-}{2}\right) = 1 - \frac{-\frac{1}{2}+1}{4} = 1 - \frac{1}{8} = \frac{7}{8}$$

$$P(X = 0) = F_X(0) - F_X\left(0^-\right) = \frac{3}{4} - \frac{1}{4} = \frac{1}{2}$$

➢ **Example 2.4:** The following is the probability mass function (PMF) of a RV X:

x	2	4	6	8
$P(X = x)$	0.3	0.2	0.4	0.1

Find the CDF $F_X(x)$ of the RV X.

- **Solution:** We know that $F_X(x) = P(X \le x) = \sum_{i:x_i \le x} P(X = x_i)$. Hence, the CDF of X is:

$$F_X(x) = \begin{cases} 0, x < 2, \\ 0.3, 2 \le x < 4, \\ 0.5, 4 \le x < 6, \\ 0.9, 6 \le x < 8, \\ 1, x \ge 8. \end{cases}$$

- **Example 2.5:** The following is the PMF of a RV X:

$$P(X = x) = \begin{cases} \frac{x}{6}, x = 1,2,3, \\ 0, \text{otherwise}. \end{cases}$$

Find the CDF $F_X(x)$ of the RV X.

- **Solution:** We know that $F_X(x) = P(X \le x) = \sum_{i:x_i \le x} P(X = x_i)$. Hence, the CDF of X is (Figure 2.1):

$$F_X(x) = \begin{cases} 0, x < 1, \\ \frac{1}{6}, 1 \le x < 2, \\ \frac{1}{2}, 2 \le x < 3, \\ 1, x \ge 3. \end{cases}$$

2.2.1 DISCRETE TYPE RANDOM VARIABLE (RV)

We observed that F_X may be continuous everywhere in $\mathbb{R}$ or it may have discontinuities in $\mathbb{R}$. If the sum of sizes of jumps at the point of discontinuities of F_X is equal to one, then the associated RV X is of discrete type.

- **Definition 2.3:** Let X be a RV with CDF F_X, and let S_X denote the set of discontinuity points of F_X. Suppose that $J_X(x) = F_X(x) - F_X(x-)$ denote the size of the jump at the discontinuity point x, where

$F(x-) = \lim_{h \to 0} F(x-h)$. The RV X is said to be of discrete type if $\sum_{x \in S_X} p_X(x) = 1$. The set S_X is called the support of the RV X. The function p_X: $\mathbb{R} \to \mathbb{R}$,

$$p_X(x) = \begin{cases} J_X(x), x \in S_X \\ 0, x \notin S_X, \end{cases}$$

is called the PMF of the discrete RV X.

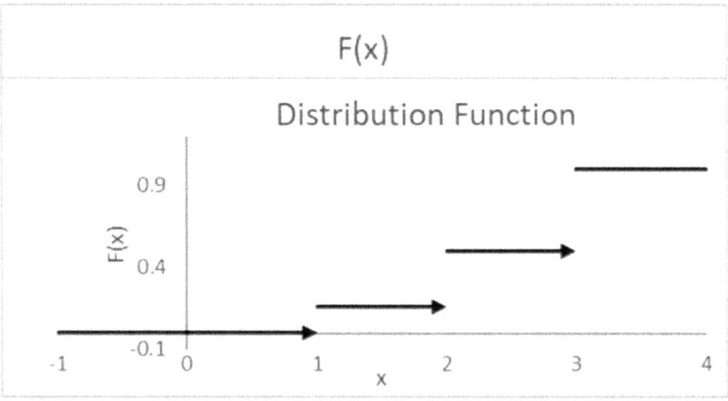

FIGURE 2.1 Distribution function of a RV X

The PMF of a discrete RV must satisfy the following two conditions:

i. $p_X(x) \geq 0$, $\forall x \in \mathbb{R}$; and
ii. $\sum_{x \in S_X} p_X(x) = 1$.

Note that if a function $g(x)$ satisfies the above two conditions, then $g(x)$ is a PMF of some discrete RV.

Note that the function $P_X(x)$, given in Example 2.2, satisfy the above properties of the PMF of a RV. Hence, the function $P_X(x)$ in Example 2.2 is the PMF of the RV X, which represents the points of the grade awarded by the institute.

➢ **Example 2.6:** A RV X has the following PMF:

x	0	1	2	3	4
$P(X = x)$	0.1	a	0.2	$2a$	0.4

Find:

i. The value of a;

ii. $P(X < 3)$;
iii. $P(X \geq 3)$;
iv. $P(0 < X < 2)$;
v. Draw the graph of the PMF.

➢ **Solution:**

i. For the value of a we must verify two properties of PMF (a) $p(x) \geq 0$; and (b) $\Sigma_x p(x) = 1$. The first property gives us that constant $a \geq 0$. Now for value of a, using second property of PMF, that is:

$$\sum_x p(x) = 1$$

$$p(0) + p(1) + p(2) + p(3) + p(4) = 1$$

$$0.1 + a + 0.2 + 2a + 0.4 = 1$$

$$a = 0.1$$

For $a = 0.1$, the PMF of X is:

x	0	1	2	3	4
$P(X = x)$	0.1	0.1	0.2	0.2	0.4

ii. $P(X < 3) = \sum_{x=0}^{2} P(X = x) = 0.4$;
iii. $P(X \geq 3) = 1 - P(X < 3) = 0.6$;
iv. $P(0 < X < 2) = P(X = 1) = 0.1$
v. The graph of the PMF is given in Figure 2.2.

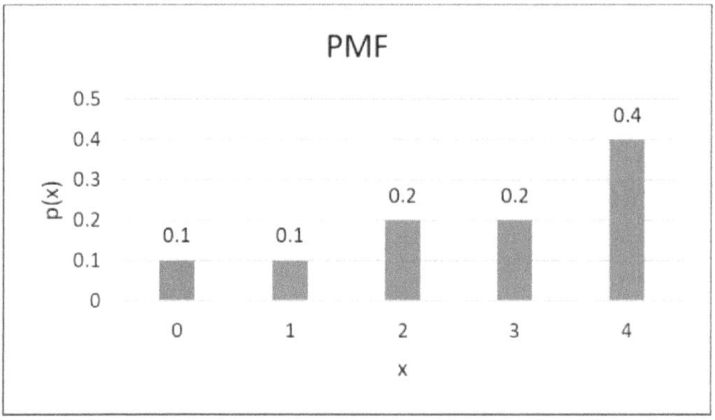

FIGURE 2.2 Graph of the PMF.

➢ **Example 2.7:** The following is the PMF of a RV *X*:

$$P(X=x)=\begin{cases}\frac{x}{10}, x=1,2,3,4,\\ 0, \text{otherwise.}\end{cases}$$

Find:
i. $P(X = 3)$;
ii. $P(1 < X < 4)$;
iii. Draw the graph of the PMF.

➢ **Solution:**
i. $P(X = 3) = 0.3$;
ii. $P(1 < X < 4); = P(X = 2 + P(X = 3) = 0.5$;
iii. The graph of the PMF is given in Figure 2.3.

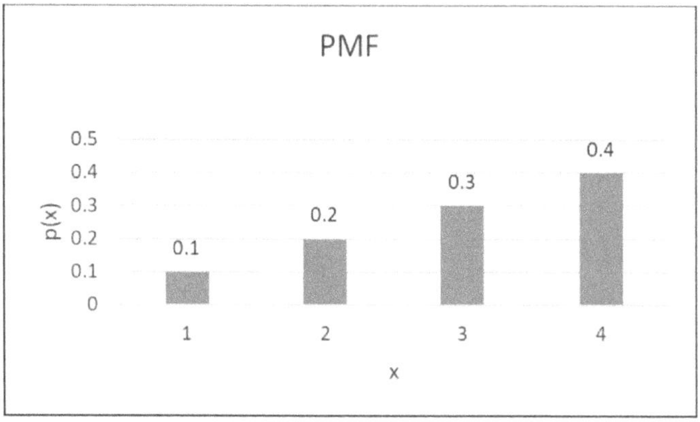

FIGURE 2.3 The PMF of the random variable X.

2.2.2 *CONTINUOUS TYPE RANDOM VARIABLE (RV)*

Recall that the CDF F_X is right continuous at every point in $\mathbb{R}$ for all types of RVs. Sometimes, we observe that for a special class of RVs, the CDF F_X is continuous everywhere in *R*, and can be written as:

$$F_X(x)=\int_{-\infty}^{x} f_X(t)\,dt, x\in\mathbb{R}$$

where; $f_X(\cdot)$ is a real-valued, non-negative, and integrable function defined on $\mathbb{R}$. The function f_X is called the probability density function (PDF), and the associated RV X is called the continuous RV. Note that the set $S_X = \{x \in \mathbb{R} : f_X(x) > 0\}$ is called the support of the continuous RV X.

The PDF of a continuous RV must satisfy the following two conditions:

i. $f_X(x) \geq 0, \forall x \in \mathbb{R}$;

ii. $\int_{-\infty}^{\infty} f_X(x)\,dx = 1$.

Note that if a function $g(x)$ satisfies the above two conditions, then $g(x)$ is a PDF of some continuous RV.

➢ **Remark:** Let a and b are real constants. The following probability can be calculated, by using the CDF of continuous RV X, as follows:

$$P(a \leq X \leq b) = P(a < X \leq b) = P(a \leq X < b)$$
$$= P(a < X < b) = F_X(b) - F_X(a).$$

➢ **Example 2.8:** The CDF of a RV X is given by:

$$F_X(x) = \begin{cases} 1 - e^{-x}, x \geq 0, \\ 0, x < 0. \end{cases}$$

Find:

i. $P(1 \leq X \leq 2)$;

ii. $P(1 < X \leq 2)$;

iii. $P(1 \leq X < 2)$;

iv. $P(1 < X < 2)$;

v. $P(X = 1)$.

➢ **Solution:** Since continuous RV, it follows those parts (i)–(iv) give the equal probabilities, and the required probability is given by:

$$F_X(2) - F_X(1) = 1 - e^{-2} - 1 + e^{-1} = e^{-1} - e^{-2}$$

The probability given in part (v) is:

$$P(X = 1) = F_X(1) - F_X(1^-) = 1 - e^{-1} - (1 - e^{-1}) = 0$$

➢ **Remark:** Note that if X is a continuous type RV, then $P(X = c) = 0$ of any constant $c \in \mathbb{R}$. Thus $P(X = 1) = 0$ in part (v) of the above example.

➢ **Example 2.9:** The following is the PDF of a continuous RV:

$$f(x) = \begin{cases} \frac{3}{4}x(2 - x), 0 < x < 2, \\ 0, \text{otherwise.} \end{cases}$$

Find the CDF $F_X(x)$ and draw its graph.

➢ **Solution:** We know that:

$$F_X(x) = P(X \le x)$$

$$= \int_{-\infty}^{x} f(t)dt$$

$$= \int_{0}^{x} \frac{3}{4} t(2-t)dt$$

$$= \frac{3}{4}\left[t^2 - \frac{t^3}{3}\right]_0^x$$

Hence, the CDF of X is:

$$F_X(x) = \begin{cases} 0, x < 0, \\ \frac{1}{4}(3x^2 - x^3), 0 \le x < 2, \\ 1, x \ge 2. \end{cases}$$

The graph of the CDF is given in Figure 2.4.

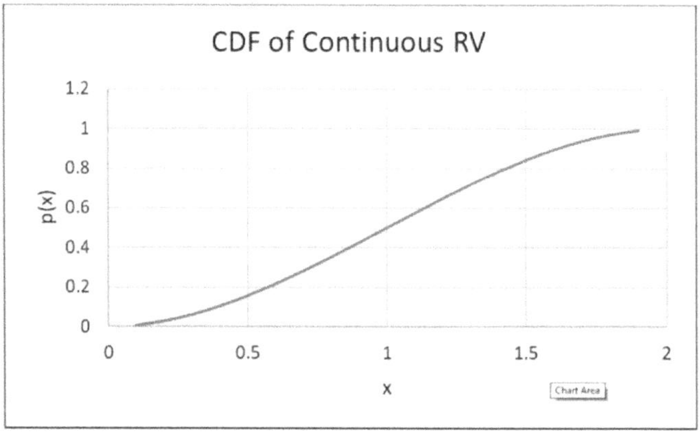

FIGURE 2.4 CDF of continuous RV.

➢ **Example 2.10:** The following is the PDF of a continuous RV:

$$f(x) = \begin{cases} 0.04x, 0 < x < 5, \\ 0.04(10-x), 5 < x < 10, \\ 0, \text{otherwise}. \end{cases}$$

Find the CDF $F_X(x)$.

- **Solution:** We know that $F_X(x) = \int_{-\infty}^{x} f(t)dt$. The CDF $F_X(x)$ is:

$$F_X(x) = \begin{cases} 0, x < 0, \\ 0.02x^2, 0 \le x < 5, \\ 0.02(20x - x^2) - 1, 5 \le x < 10, \\ 1, x \ge 10. \end{cases}$$

- **Example 2.11:** A RV *X* has the following PDF:

$$f(x) = \begin{cases} ax(1-x), 0 < x < 1, \\ 0, \text{otherwise.} \end{cases}$$

Find the value of *a* for which *f*(*x*) is PDF and draw the graph of the PDF.

- **Solution:** For the value of *a* we must verify two properties of pmf (i) $f(x) \ge 0, \forall x$, which gives us that $a \ge 0$ and by using (ii) $\int_x f(x)dx = 1$, we can obtain the value of constant *a*. Now:

$$\int_x f(x)dx = 1$$

$$\int_0^1 ax(1-x)dx = 1$$

$$a\left[\frac{x^2}{2} - \frac{x^3}{3}\right]_0^1 = 1$$

$$a = 6.$$

Hence, for $a=6$, *f*(*x*) is PDF. The graph of the PDF is given in Figure 2.5.

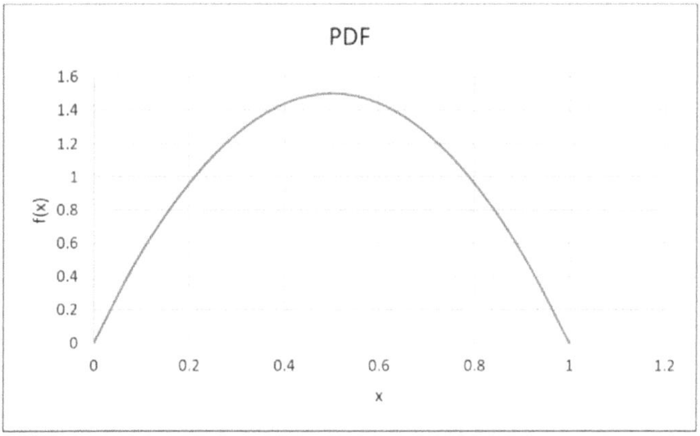

FIGURE 2.5 Graph of the PDF.

➢ **Example 2.12:** A RV *X* has the following PDF:

$$f(x)=\begin{cases}0.04x, 0<x<5,\\ 0.04(10-x), 5<x<10,\\ 0, \text{otherwise.}\end{cases}$$

Find *P*(*X* > 7).

➢ **Solution:**

$$P(X<7)=\int_0^7 f(x)dx$$

$$=\int_0^5 f(x)dx+\int_5^7 f(x)dx$$

$$=\int_0^5 0.04x dx+\int_5^7 0.04(10-x)dx$$

$$=0.04\left[\frac{x^2}{2}\right]_0^5+0.04\left[10x-\frac{x^2}{2}\right]_5^7$$

$$=0.82$$

➢ **Example 2.13:** The expression of the PDF is:

$$f_X(x)=\begin{cases}\lambda e^{-\lambda x}, 0<x<\infty,\\ 0, \text{otherwise.}\end{cases}$$

Find the CDF and draw its graph when $\lambda = 1$ and 5.

➢ **Solution:** We know that:

$$F_X(x)=\int_0^x f(t)dt$$

$$=\int_0^x \lambda e^{-\lambda t}\,dt$$

$$=1-e^{-\lambda x}$$

$$F_X(x)=\begin{cases}0, x<0,\\ 1-e^{-\lambda x}, x\geq 0.\end{cases}$$

The graph of CDF when $\lambda = 1$ is given in Figure 2.6.

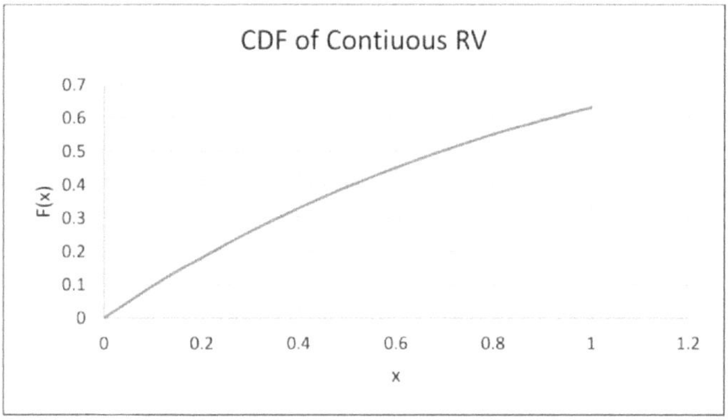

FIGURE 2.6 Graph of CDF when $\lambda = 1$.

The graph of CDF when $\lambda = 5$ is given in Figure 2.7.

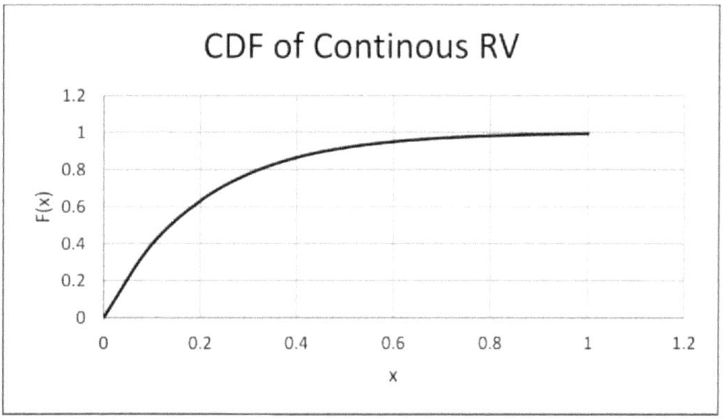

FIGURE 2.7 Graph of CDF when $\lambda = 5$.

2.3 EXPECTATION OF RANDOM VARIABLE (RV)

The characteristic of a probability model can be described by its moments. Let X be a RV having a probability distribution. The expectation of RV is defined as:

$$E(X) = \begin{cases} \int_x xf(x)dx, \text{if } X \text{ is a continuous RV,} \\ \sum_x xp(x), \text{if } X \text{ is a discrete RV,} \end{cases}$$

Provided $\int_x |x| f(x)dx$ and $\sum_x |x| p(x)$ are exists.

Let $g(X)$ be a measurable function of RV X. The expectation of a function of the RV is defined as:

$$E(g(X)) = \begin{cases} \int_x g(x)f(x)dx, \text{if } X \text{ a is continuous RV,} \\ \sum_x g(x)p(x), \text{if } X \text{ is a discrete RV,} \end{cases}$$

provided $\int_x | g(x)| f(x)dx$ and $\sum_x | g(x)| p(x)$ are exists.

2.4 MOMENTS OF RANDOM VARIABLE (RV)

The rth raw moments or moments about the origin is defined as:

$$\mu_r' = E\left(X^r\right), r = 1,2,\ldots.$$

The first four moments are useful to analyze the shape (or behavior) of the probability models. The first moment $\mu_1' = E\left(X\right) = \mu$ (say) is known as the mean (or the population mean) of the distribution. Further, the rth central moments (or moments about the mean) is defined as:

$$\mu_r = E\left(X - \mu\right)^r, r = 1,2,\ldots.$$

Here; μ is the mean of the distribution. The first central moment is always zero for every RV, i.e., $\mu_1 = E\left(X - \mu\right) = 0$. The second central moment is $\mu_2 = E\left(X - \mu\right)^2 = E\left(X^2\right) - \mu^2$. It is called the variance of the RV X and also known as the measure of spread, i.e., $\mu_2 = \sigma^2 = V\left(X\right) = E\left(X - \mu\right)^2$. The third and fourth central moments are useful in measuring the skewness and kurtosis, respectively. For measuring the skewness and kurtosis, Karl Pearson defined the following four coefficients based upon the first four central moments.

Measure	**Formula**	
Skewness	$\beta_1 = \frac{\mu_3^2}{\mu_2^3}$	$\gamma_1 = \sqrt{\beta_1} = \frac{\mu_3}{\left(\mu_2\right)^{3/2}}$
Kurtosis	$\beta_2 = \frac{\mu_4}{\mu_2^2}$	$\gamma_2 = \beta_2 - 3$

For symmetrical distribution, β_1 shall be zero. The value of $\beta_1 > 0$ indicates the skewness of the distribution but β_1 does not provide the direction of the skewness, i.e., positive skewed or negative skewed. So, we can use another coefficient of skewness γ_1, which provides the direction of the skewness. Since μ_2 is always positive, it follows that the sign of skewness would depend on the sign of μ_3. Similarly, the values of β_2 and γ_2 measure of the kurtosis If $\gamma_2 = 0, \gamma_2 > 0, \gamma_2 < 0$, the distribution is normal (or mesokurtic), leptokurtic, and platykurtic distribution, respectively.

- **Properties of Expectation:**
  1. Let X be a non-negative RV then $E(X) \geq 0$, provided $E(X)$ exists.
  2. Let X and Y be two RVs such that $E(X)$ and $E(Y)$ are exists. Let a, b and c are three non-zero real constants, then:

     $$E(aX + bY + c) = aE(X) + bE(Y) + c.$$

  3. Let X be a RV and $g(X)$ is a function of RV X, then:

     $$E(ag(X) + b) = aE(g(X)) + b$$

     where; a and b are real constant.
  4. Let X and Y are independent RVs such that $E(X)$ and $E(Y)$ exists then:

     $$E(XY) = E(X)E(Y).$$

  5. The expected value of constant is constant. That is, for a given constant c:

     $$E(c) = c.$$

  6. If X and Y are independent RVs with a condition that $P(X \leq Y) = 1$, then:

     $$E(X) \leq E(Y).$$

  7. Let $E(X^r)$ exists. If we take $0 < s \leq r$, then $E(X^S)$ also exists.

- **Properties of Variance:**
  1. A variance of constant is zero. That is:

     $$V(c) = 0.$$

  2. Let X be a RV and a, b are real constants then:

     $$V(aX + b) = a^2 V(X).$$

     That is, the variance is independent of location b but not scale a.

3. Let X and Y be two independent RVs then:

$$V(X+Y)=V(X)+V(Y).$$

- **Example 2.14:** Find the $E(X)$ and $E(X^2)$ for the PDF given in Example 2.13.
- **Solution:** Mean of RV X is:

$$E(X)=\int_x xf(x)dx$$

$$=\lambda\int_0^\infty xe^{-\lambda x}\,dx$$

$$=\frac{1}{\lambda}$$

Now, the second moment about the origin is:

$$E(X^2)=\int_x x^2 f(x)dx$$

$$=\int_0^\infty \lambda x^2 e^{-\lambda x}\,dx$$

$$=\frac{2}{\lambda^2}$$

- **Example 2.15:** A RV X has the following PDF:

$$f(x)=\begin{cases}3x^2, 0<x<1,\\ 0, \text{otherwise.}\end{cases}$$

Find the $E(X)$ and $V(X)$.

- **Solution:** Mean of RV X is:

$$E(X)=\int_x xf(x)dx=3\int_0^1 x^3\,dx=\frac{3}{4}$$

Similarly, the second moment about the origin is:

$$E(X^2)=\int_0^1 x^2 f(x)dx=3\int_0^1 x^4\,dx=\frac{3}{5}$$

Now, the variance of RV X is:

$$V(X)=E(X^2)-\left(E(X)\right)^2=\frac{3}{5}-\frac{9}{16}=\frac{3}{80}$$

2.5 MOMENT GENERATING FUNCTION (MGF)

The moment generating function (MGF) of X gives us all moments of RV X (if MGF exists). That is why it is called the MGF. If MGF exists, it uniquely determines the distribution. If two RVs have the same MGFs, then they must have the same distributions. Thus, if you find the MGF of a RV, you have indeed determined its distribution. The MGF of a RV X is defined as:

$$M_X(t) = \begin{cases} \int_x e^{tx} f(x)dx, \text{if } X \text{ is a continuous RV,} \\ \sum_x e^{tx} p(x), \text{if } X \text{ is a discrete RV,} \end{cases}$$

where $|t| < h, \text{for some } h > 0$

Now, consider:

$$M_X(t) = E\left(e^{tx}\right)$$

$$= E\left(\sum_{r=0}^{\infty} \frac{(tx)^r}{r!}\right)$$

$$= \sum_{r=0}^{\infty} E(X^r)\frac{t^r}{r!}$$

$$M_X(t) = \sum_{r=0}^{\infty} \mu_r' \frac{t^r}{r!}$$

That is, coefficient of $\frac{t^r}{r!}$ $M_X(t)$ gives r^{th} raw moments (or moments about origin). Therefore:

$$\mu_r' = E\left(X^r\right) = \text{coefficient of } \frac{t^r}{r!} \text{in } M_X(t).$$

- **Result:** We can also find, r^{th} raw moments μ_r' using the following formula:

$$\mu_r' = E\left(X^r\right) = \frac{\partial^r}{\partial t^r} M_X(t)\bigg|_{t=0}, r = 1, 2, \ldots$$

- **Properties of Moment Generating Function:**
  1. Le c be non-zero real constant then:

$$M_{cX}(t) = M_X(ct)$$

  2. Let $X_1, X_2, \ldots X_n$, be independent RVs. Define $Y = \sum_{i=1}^{n} X_i$, then MGF of Y is:

$$M_Y(t) = M_{\sum_{i=1}^{n} X_i}(t) = \prod_{i=1}^{n} M_{X_i}(t)$$

3. Let X be a RV having MGF $M_X(t)$. Consider $Y = \frac{X-a}{b}$ for $b \neq 0$. Then, the MGF of Y is:

$$M_Y(t) = e^{-\frac{at}{b}} M_X\left(\frac{t}{b}\right).$$

4. The MGF of the distribution, if it exists, uniquely determines the distribution, that is:

$M_X(t) = M_Y(t) \Rightarrow X \& Y$ have the same CDFs.

- **Example 2.16:** Find MGF expression and find four raw moments about origin using the PDF given in Example 2.13.
- **Solution:** We know that:

$$M_X(t) = \int_x e^{tx} f(x)dx$$

$$= \int_0^{\infty} e^{tx} \lambda e^{-\lambda x} dx$$

$$= \lambda \int_0^{\infty} e^{-(\lambda - t)x} dx$$

$$M_X(t) = \left(1 - \frac{t}{\lambda}\right)^{-1}, t < \lambda$$

We can find raw moments using MGF.

The coefficient of $\frac{t^r}{r!}$ $M_X(t)$ gives r^{th} raw moments (or moments about origin):

$$M_X(t) = \left(1 - \frac{t}{\lambda}\right)^{-1}, t > \lambda$$

Using the expansion $(1-a)^{-1} = \sum_{k=0}^{8} a^k, |a| < 1,$

$$M_X(t) = \left\{1 + \left(\frac{t}{\lambda}\right) + \left(\frac{t}{\lambda}\right)^2 + \left(\frac{t}{\lambda}\right)^3 + \left(\frac{t}{\lambda}\right)^4 + \ldots\right\} = \sum_{r=0}^{\infty}\left(\frac{t}{\lambda}\right)^r = \sum_{r=0}^{\infty}\frac{r!}{\lambda^r}\frac{t^r}{r!}$$

μ'_r = coefficient of $\frac{t^r}{r!}$ in $M_X(t) = \frac{r!}{\lambda^r}; r = 1, 2, \ldots$

Thus,

$$\mu'_1 = \frac{1}{\lambda}, \mu'_2 = \frac{2}{\lambda^2}, \mu'_3 = \frac{6}{\lambda^3}, \mu'_4 = \frac{24}{\lambda^4}$$

- **Alternative Method:** We can also find, r^{th} raw moments μ_r' using the following formula:

$$\mu_r' = E\left(X^r\right) = \frac{\partial^r}{\partial t^r} M_X(t)\bigg|_{t=0} \quad ;r = 1, 2, \ldots,$$

Substitute $r = 1, 2$ and solve them; we can get the first and second raw moments μ_1', μ_2',

$$\mu_1' = \frac{\partial}{\partial t} M_X(t)\bigg|_{t=0}$$

Now,

$$\frac{\partial}{\partial t} M_X(t) = \frac{\partial}{\partial t}\left(1 - \frac{t}{\lambda}\right)^{-1} = \frac{1}{\lambda}\left(1 - \frac{t}{\lambda}\right)^{-2}, t > \lambda$$

Hence,

$$\mu_1' = \frac{\partial}{\partial t} M_X(t)\bigg|_{t=0} = \frac{1}{\lambda}$$

Further, the second raw moment:

$$\mu_2' = \frac{\partial^2}{\partial t^2} M_X(t)\bigg|_{t=0}$$

Now,

$$\frac{\partial^2}{\partial t^2} M_X(t) = \frac{\partial}{\partial t}\left[\frac{\partial}{\partial t} M_X(t)\right]$$

$$= \frac{\partial}{\partial t}\left[\frac{1}{\lambda}\left(1 - \frac{t}{\lambda}\right)^{-2}\right]$$

$$= \frac{2}{\lambda^2}\left(1 - \frac{t}{\lambda}\right)^{-3}, t > \lambda$$

$$\mu_2' = \frac{\partial^2}{\partial t^2} M_X(t)\bigg|_{t=0} = \frac{2}{\lambda^2}$$

Similarly, we can find the third and fourth raw moments.

2.6 CUMULANT GENERATING FUNCTION (CGF)

Cumulants of a probability distribution is a sequence of numbers that compactly define the distribution. The cumulant generating function (CGF) of a RV X is defined as:

$$\kappa_X(t) = \log M_X(t)$$

Now, the right-hand side can be expanded as a convergent series in powers of t. Thus,

$$\kappa_X(t) = \sum_{r=0}^{\infty} \kappa_r \frac{t^r}{r!}$$

That is, the coefficient of $\frac{t^r}{r!}$ $K_X(t)$ gives r^{th} cumulants. Therefore:

$$\kappa_r = \text{coefficient of } \frac{t^r}{r!} \text{ in } \kappa_X(t)$$

- **Result:** We can also find, r^{th} cumulants κ_r using the following formula:

$$\kappa_r = \frac{\partial^r}{\partial t^r} \kappa_X(t)\Big|_{t=0} \quad ; r = 1, 2, \ldots,$$

- **Result:** The relationship between cumulants and moments is given below:

$$\mu_1' = \kappa_1$$
$$\mu_2 = \kappa_2$$
$$\mu_3 = \kappa_3$$
$$\mu_4 = 3\kappa_2^2 + \kappa_4$$

- **Example 2.17:** Find the expression of the CGF and find mean, second, third, and fourth central moments using the PDF given in Example 2.13.
- **Solution:** We know that:

$$\kappa_X(t) = log\,(M_X(t))$$

Substituting the expression of MGF from the previous example, we get:

$$\kappa_X(t) = \log\left(1 - \frac{t}{\lambda}\right)^{-1}, t < \lambda$$

We can find central moments using CGF as follows:

$$\kappa_r = \text{coefficient of } \frac{t^r}{r!} \text{ in } \kappa_X(t)$$

Therefore,

$$\kappa_X(t) = log\left(1 - \frac{t}{\lambda}\right)^{-1}$$

$$= -log\left(1-\frac{t}{\lambda}\right), t < \lambda$$

Using $log(1-x) = -x - \frac{x^2}{2} - \frac{x^3}{3} - \cdots, x < 1$ expansion, we get:

$$= -\left\{-\frac{1}{1}\frac{t}{\lambda} - \frac{1}{2}\left(\frac{t}{\lambda}\right)^2 - \frac{1}{3}\left(\frac{t}{\lambda}\right)^3 - \frac{1}{4}\left(\frac{t}{\lambda}\right)^4 + \cdots\right\}$$

$$= \frac{1}{1!}\frac{t}{\lambda} + \frac{1}{2!}\left(\frac{t}{\lambda}\right)^2 + \frac{2}{3!}\left(\frac{t}{\lambda}\right)^3 + \frac{6}{4!}\left(\frac{t}{\lambda}\right)^4 + \cdots$$

$\kappa_1 = \mu_1'$, = mean= coefficient of $\frac{t}{1!}$ in $\kappa_X(t) = \frac{1}{\lambda}$

$\kappa_2 = \mu_2$ = variance = coefficient of $\frac{t^2}{2!}$ in $\kappa_X(t) = \frac{1}{\lambda^2}$

$\kappa_3 = \mu_3$ = coefficient of $\frac{t^3}{3!}$ in $\kappa_X(t) = \frac{2}{\lambda^3}$

κ_4 = coefficient of $\frac{t^4}{4!}$ in $\kappa_X(t) = \frac{6}{\lambda^4}$

$$\mu_4 = \kappa_4 + 3\kappa_2^2 = \frac{6}{\lambda^4} + 3\left(\frac{1}{\lambda^2}\right)^2 = \frac{9}{\lambda^4}$$

- **Alternative Method:** We can also find, r^{th} cumulants using the following formula:

$$\kappa_r = \frac{\partial^r}{\partial t^r}\kappa_X(t)\Big|_{t=0}, r = 1, 2, \ldots$$

Substitute r = 1, 23 and 4 and solve them; we can get κ_1, κ_2, κ_3, and κ_4

$$\kappa_1 = \frac{\partial}{\partial t}\kappa_X(t)\Big|_{t=0}$$

Now,

$$\frac{\partial}{\partial t}\kappa_X(t) = \frac{\partial}{\partial t}log\left(1-\frac{t}{\lambda}\right)^{-1} = \frac{1}{\left(1-\frac{t}{\lambda}\right)^{-1}}\frac{1}{\lambda}\left(1-\frac{t}{\lambda}\right)^{-2}$$

Hence,

$$\mu_1' = \frac{\partial}{\partial t}\kappa_X(t)\Big|_{t=0} = \frac{1}{\lambda}$$

Further, the second raw moment:

$$\mu_2 = \frac{\partial^2}{\partial t^2}\kappa_X(t)\Big|_{t=0}$$

Now,

$$\frac{\partial^2}{\partial t^2}\kappa_X(t) = \frac{\partial}{\partial t}\left[\frac{\partial}{\partial t}\kappa_X(t)\right]$$

$$= \frac{\partial}{\partial t}\left[\frac{1}{\lambda}\left(1-\frac{t}{\lambda}\right)^{-1}\right]$$

$$= \frac{1}{\lambda^2}\left(1-\frac{t}{\lambda}\right)^{-2}$$

$$\mu_2 = \frac{\partial^2}{\partial t^2}\kappa_X(t)\Big|_{t=0} = \frac{1}{\lambda^2}.$$

Similarly, we can find the third and fourth central moments.

- **Example 2.18:** Find β_1, γ_1, β_2, γ_2 for the PDF given in Example 2.13. Also, give the interpretation.
- **Solution:** Using the results second, third, and fourth central moments from the above example, we get:

$$\beta_1 = \frac{\mu_3^2}{\mu_2^3} = \frac{\left(\frac{2}{\lambda^3}\right)^2}{\left(\frac{1}{\lambda^2}\right)^3} = 4$$

$$\gamma_1 = \frac{\mu_3}{\mu_2^{3/2}} = 2$$

$$\beta_2 = \frac{\mu_4}{\mu_2^2} = \frac{\frac{9}{\lambda^4}}{\left[\frac{1}{\lambda^2}\right]^2} = 9$$

$$\gamma_2 = 9 - 3 = 6$$

This implies that the given probability model is positively skewed and leptokurtic.

2.7 CHARACTERISTIC FUNCTION (CF)

The advantage of the characteristic function (CF) is that it always exists, unlike the MGF. The CF of a RV X is a function $\varphi_X(t)$ defined as:

$$\phi_X(t) = \begin{cases} \int_x e^{itx} f(x)dx, \text{if } X \text{ is a continuous RV,} \\ \sum_x e^{itx} p(x), \text{if} X \text{ is a discrete RV.} \end{cases}$$

- **Example 2.19:** Find the CF of the PDF given in Example 2.13.
- **Solution:** We know that:

$$\varphi_X(t) = \int_x e^{itx} f(x)dx$$

$$= \int_0^{\infty} e^{itx} \lambda e^{-\lambda x}\, dx$$

$$= \lambda \int_0^{\infty} e^{-(\lambda - it)x}\, dx$$

$$\phi_X(t) = \left(1 - \frac{it}{\lambda}\right)^{-1}.$$

KEYWORDS

- **cumulative distribution function**
- **cumulative performance index**
- **positively skewed and leptokurtic**
- **probability density function**
- **probability mass function**
- **random variable**

REFERENCES

Casella, G., & Berger, R. L., (2002). *Statistical Inference*. Duxbury: Belmont, CA.

Gupta, S. C., & Kapoor, V. K., (1997). *Fundamentals of Mathematical Statistics*. Sultan Chand and Sons: New Delhi.

Heumann, C., Schomaker, M., & Shalabh, (2016). *Introduction to Statistics and Data Analysis: With Exercises, Solutions and Applications in R*. Springer.

Hogg, R. V., McKean, J. W., & Craig, A. T., (2013). *Introduction to Mathematical Statistics* (7th edn.). Pearson: Boston.

Miller, I., Miller, M., Freund, J. E., & Miller, I., (2004). *John E. Freund's Mathematical Statistics with Applications*. Prentice-Hall: Upper Saddle River, NJ.

Rohatgi, V., & Saleh, A. K. Md. E., (2015). *Introduction to Probability and Statistics* (3rd edn.). John Wiley.

CHAPTER 3

Discrete Distributions

MUKTI KHETAN[1], VIJAY KUMAR[2], MOHD. ARSHAD[3], and PRASHANT VERMA[4]

[1]*Department of Mathematics, Amity School of Applied Sciences, Amity University Mumbai, Maharashtra, India*

[2]*Department of Mathematics, Indian Institute of Technology Bombay, Powai, Mumbai, Maharashtra, India*

[3]*Department of Mathematics, Indian Institute of Technology Indore, Simrol, Indore, Madhya Pradesh, India*

[4]*Department of Statistics, Faculty of Science, University of Allahabad, Prayagraj, India*

3.1 UNIFORM DISTRIBUTION (DISCRETE)

The uniform distribution contains a fixed number of values $(\kappa_1, \kappa_2, \ldots, \kappa_l)$ with probability $(1/l)$. The probability model wherein a finite (predetermine) number of values 'l' are equally likely to be observed; each value of 'l' has the same probability $1/l$. In uniform distribution, the expression of probability mass function (PMF) of the random variable (RV) X is:

$$P(X = x) = \begin{cases} \frac{1}{l} & x = \kappa_1, \kappa_2, \ldots, \kappa_l \\ 0 & \text{otherwise} \end{cases},$$

Distribution Theory: Principles and Applications. Fozia Homa, Mukti Khetan, Mohd. Arshad, & Pradeep Mishra (Naz) (Eds.)

where; $\kappa_1, \kappa_2, \ldots, \kappa_l$ are real constant. We can use $\kappa_l = l \;\; \forall l = 1, 2, ..., n$. In this case, we define a discrete uniform distribution over the set $\{1, 2,\ldots,n\}$ as:

$$P(X = x) = \begin{cases} \dfrac{1}{n} & x = 1, 2, ..., n \\ 0 & \text{otherwise} \end{cases}$$

Symbolically: X~Uniform(n).
Here n is a parameter of the distribution.

- **Sample Generation from Uniform distribution:** For sample generation of discrete uniform distribution (Figure 3.1), we must install one package using the following coding:

```
install.package("purr")
library(purr)
```

R Code:

$x = rdunif(n, b, a)$
$where$
$n = sample\ size$
$a, b = \ min\ and\ max\ value$
$Example$
$x = rdunif(10\ , 0, 1)$

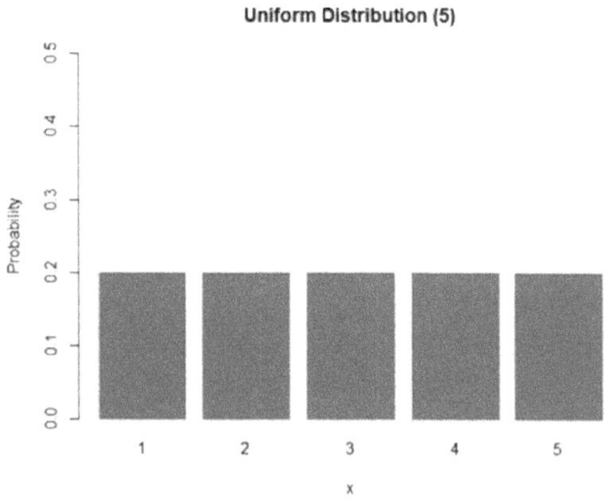

FIGURE 3.1 Uniform distribution.

- **Output:**

```
Console  Terminal ×  Jobs ×
~/
> library(purrr)
> x=rdunif(10 , 0, 1)
> x
[1] 1 1 1 1 0 1 1 0 0 1
```

3.1.1 PROPERTIES OF THE UNIFORM DISTRIBUTION

a. **Raw Moments of Uniform Distribution:** If X is the random variable (RV), the rth raw moment is defined below:

$$\mu_r' = E(X^r) = \sum_x x^r P(X = x).$$

The expression of the μ_1' is:

$$\mu_1' = E(X) = \sum_{x=1}^{n} xP(X = x) = \sum_{x=1}^{n} x\frac{1}{n}.$$

Since $\sum_{i=1}^{n} i = \frac{1}{2}n(n+1)$, we get:

$$\Rightarrow \mu_1' = \frac{(n+1)}{2}. \qquad (3.1.1)$$

Thus, $\frac{(n+1)}{2}$ is the mean of the distribution.

The expression of the μ_2' is:

$$\mu_2' = E(X^2) = \sum_{x=1}^{n} x^2 P(X = x) = \sum_{x=1}^{n} x^2 \frac{1}{n}.$$

Since $\sum_{i=1}^{n} i^2 = \frac{1}{6}n(2n+1)(n+1)$, we get:

$$\Rightarrow \mu_2' = \frac{(2n+1)(n+1)}{6}. \qquad (3.1.2)$$

The expression of the μ_3' is:

$$\mu_3' = E(X^3) = \sum_{x=1}^{n} x^3 P(X = x) = \sum_{x=1}^{n} x^3 \frac{1}{n}.$$

Since $\sum_{i=1}^{n} i^3 = \left[\frac{1}{2}n(n+1)\right]^2$, we get:

$$\Rightarrow \mu_3' = \frac{n(n+1)^2}{4}. \qquad (3.1.3)$$

The expression of the μ_4' is:

$$\mu_4' = E(X^4) = \sum_{x=1}^{n} x^4 P(X = x) = \sum_{x=1}^{n} x^4 \frac{1}{n}.$$

Since $\sum_{i=1}^{n} i^4 = \frac{1}{30}n(n+1)(6n^3 + 9n^2 + n - 1)$ we get:

$$\Rightarrow \mu_4' = \frac{1}{30}(n+1)(6n^3+9n^2+n-1)$$

$$= \frac{1}{30}(n+1)\left[3n^2(2n+1)+6n^2+3n-2n-1\right].$$

$$\mu_4' = \frac{1}{30}(n+1)(2n+1)\left[3n^2+3n-1\right] \quad (3.1.4)$$

b. **Central Moments of Uniform Distribution:** If X is the RV, the rth central moment is defined below:

$$\mu_r = E(X-\mu)^r = \sum_x (x-\mu)^r P(X=x)$$

where; μ is the mean.

In this section, we have derived the first four central moments $\mu_1, \mu_2, \mu_3, \mu_4$. Since the first central moment is always zero, i.e.,

$$\mu_1 = 0 \quad (3.1.5)$$

The expression of the μ_2' is:

$$\mu_2 = \mu_2' - \mu_1'^2$$

$$= \frac{(2n+1)(n+1)}{6} - \left[\frac{(n+1)}{2}\right]^2 \text{ [using Eqns. (3.1.1) and (3.1.2)]}$$

$$= \frac{(n+1)}{2}\left[\frac{(2n+1)}{3} - \frac{(n+1)}{2}\right]$$

$$= \frac{(n^2-1)}{12}. \quad (3.1.6)$$

Thus, the *variance* of the distribution as mentioned above is $\frac{(n^2-1)}{12}$.

The expression of the μ_3' is:

$$\mu_3 = \mu_3' - 3\,\mu_2'\mu_1' + 2\mu_1'^3$$

$$= \frac{n(n+1)^2}{4} - 3\frac{(2n+1)(n+1)}{6}\frac{(n+1)}{2} + 2\left[\frac{(n+1)}{2}\right]^3$$

[using Eqns. (3.1.1)–(3.1.3)]

$$= 0 \quad (3.1.7)$$

The expression of the μ_4' is:

$$\mu_4 = \mu_4' - 4\,\mu_3'\,\mu_1' + 6\,\mu_2'\,\mu_1'^2 - 3\mu_1'^4$$

$$= \frac{1}{30}(n+1)(2n+1)\left[3n^2+3n-1\right] - 4\frac{n(n+1)^2}{4}\frac{(n+1)}{2}$$

$$+ 6\frac{(2n+1)(n+1)}{6}\left[\frac{(n+1)}{2}\right]^2 - 3\left[\frac{(n+1)}{2}\right]^4$$

[using Eqns. (3.1.1)–(3.1.4)]

$$= \frac{1}{30}(n+1)(2n+1)\left[3n^2+3n-1\right] + \frac{(n+1)^3}{2}\left[-n+\frac{(2n+1)}{2}-\frac{3(n+1)}{8}\right]$$

$$= \frac{1}{240}(n+1)\left[48n^3+72n^2+8n-8+15(n^2+2n+1-3n^3-6n^2-3n)\right]$$

$$= \frac{1}{240}(n+1)\left[3n^3-3n^2-7n+7\right]$$

$$= \frac{1}{240}(n+1)\left[3n^2(n-1)-7(n-1)\right]$$

$$= \frac{1}{240}(n^2-1)(3n^2-7) \qquad (3.1.8)$$

c. **Skewness and Kurtosis of Uniform Distribution:** Karl Pearson defined the four coefficients for skewness and kurtosis based upon the first four central moments. The expressions of skewness (β_1, γ_1) and kurtosis (β_2, γ_2) are given below:

$$\beta_1 = \frac{\mu_3^2}{\mu_2^3} = 0. \text{ [using Eqns. (3.1.6) and (3.1.7)]}$$

$$\gamma_1 = \frac{\mu_3}{\mu_2^{3/2}} = 0. \qquad (3.1.9)$$

$$\beta_2 = \frac{\mu_4}{\mu_2^2} = \frac{1}{240}\frac{(n+1)(n-1)(3n^2-7)}{\left[\frac{(n^2-1)}{12}\right]^2}. \text{ [using Eqns. (3.1.6) and (3.1.8)]}$$

$$\beta_2 = \frac{3}{5}\frac{(3n^2-7)}{(n^2-1)}.$$

$$\gamma_2 = \beta_2 - 3 = \frac{3}{5}\frac{(3n^2-7)}{(n^2-1)} - 3 = -\frac{6}{5}\frac{(n^2+1)}{(n^2-1)}. \qquad (3.1.10)$$

d. Moment Generating Function (MGF) of Uniform Distribution: The MGF plays an essential role in the theory of distribution. It will make it easier to know about the characteristics of the model. Symbolically the MGF is denoted by $M_X(t)$ and is defined as:

$$M_X(t) = E(e^{tX}) = \sum_x e^{tx} P(X = x).$$

The expression for MGF $M_X(t)$ is obtained as below:

$$M_X(t) = \sum_{x=1}^{n} e^{tx} P(X = x) = \sum_{x=1}^{n} e^{tx} \frac{1}{n} = \frac{1}{n}(e^t + e^{2t} + e^{3t} + \ldots + e^{nt}).$$

As we know that the sum of the series $b + br + br^2 + \ldots + br^n$ is $\frac{b(1-r^n)}{(1-r)}$, substituting the formula in the above expression, we get:

$$\Rightarrow M_X(t) = \frac{1}{n}\left[\frac{e^t(1-e^{nt})}{(1-e^t)}\right]; \ t \neq 0, \text{ and } M_X(0)=1.$$

- **Note:** The expression probability generating function (PGF) can be derived using the same MGF method. The formula of PGF is

$$P_X(t) = \sum_x t^x P(X = x).$$

e. Characteristic Function (CF) of Uniform Distribution: The benefit of the CF is that it always exists, unlike the MGF function. Symbolically the CF is denoted by $\phi_X(t)$ and is defined as:

$$\phi_X(t) = E(e^{itX}) = \sum_x e^{itx} P(X = x).$$

The expression for CF ($\phi_X(t)$) is obtained as below:

$$\phi_X(t) = \sum_{x=1}^{n} e^{itx} P(X = x) = \sum_{x=1}^{n} e^{itx} \frac{1}{n} = \frac{1}{n}(e^{it} + e^{2it} + e^{3it} + \ldots + e^{nit}),$$

Using the formula of the sum of the series, we get:

$$\Rightarrow \phi_X(t) = \frac{1}{n}\left[\frac{e^{it}(1-e^{nt})}{(1-e^t)}\right]; t \neq 0, \text{ and } \phi_X(0) = 1.$$

3.2 BERNOULLI DISTRIBUTION

If an experiment has only two outcomes, then the probability distribution of this experiment is the *Bernoulli distribution*, which usually have zero and one with probability (1-p) and p, respectively. Here ‘p’ is the success probability and $0 \leq p \leq 1$. For example, the result of a competitive exam has two possible consequences: pass or fail, and the HIV test has two possible consequences: tested positive or negative. In Bernoulli distribution, the PMF of a random variable X is:

$$P(X = x) = \begin{cases} p^x (1\text{-}p)^{1-x} & x{=}0,1 \\ 0 & \text{otherwise} \end{cases}$$

Symbolically: X~Bernoulli(p).
Here; p is a parameter of the distribution.

➤ **Sample Generation from Bernoulli Distribution** (Figure 3.2):

R Code:

$x = rbernoulli(n, p)$
where
$n = sample\ size$
$p = Probability\ of\ true\ (success)$

Example
$x = rbernoulli(10, 0.5)$

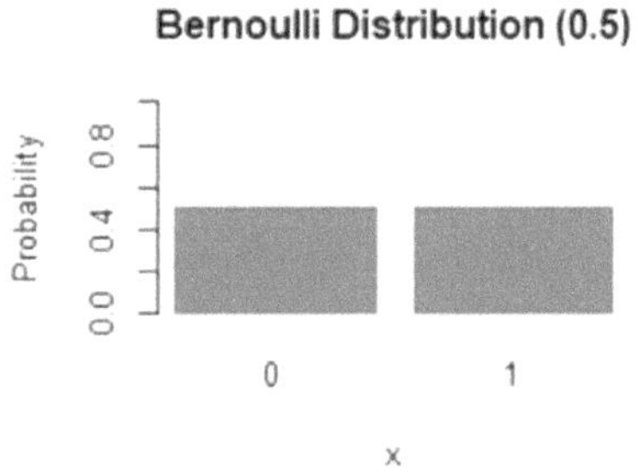

FIGURE 3.2 Bernoulli distribution.

➤ **Output:**

```
Console  Terminal  Jobs
> x=rbernoulli(10, 0.5)
> x
 [1]  TRUE FALSE FALSE FALSE FALSE  TRUE FALSE  TRUE  TRUE FALSE
```

3.2.1 PROPERTIES OF BERNOULLI DISTRIBUTION

a. **Moments of Bernoulli Distribution:** The first four raw moments μ_1', μ_2', μ_3', μ_4' have been derived as below:

$$\mu_r' = E(X^r) = \sum_{x=0}^{1} x^r P(X = x) = \sum_{x=0}^{1} x^r p^x (1\text{-}p)^{1-x} = 0p^0 (1\text{-}p)^{1-0} + 1.p^1 (1\text{-}p)^{1-1}$$

$$\mu_r' = p$$

Substitute r = 1, 2, 3, 4 in the above expression and get all fours moments $\mu_1', \mu_2', \mu_3', \mu_4'$.

$$\mu_r' = p \quad (3.2.1)$$

Thus, p is the mean of the distribution.

$$\mu_2' = p \quad (3.2.2)$$

$$\mu_3' = p \quad (3.2.3)$$

$$\mu_3' = p \quad (3.2.4)$$

b. **Central Moments Bernoulli Distribution:** The first four central moments $\mu_1', \mu_2', \mu_3', \mu_4'$ have been derived as below:

$$\mu_1 = 0 \quad (3.2.5)$$

$$\mu_2 = \mu_2' - \mu_1'^2 = p - p^2 = p(1-p) \quad \text{[using Eqns. (3.2.1) and (3.2.2)]}$$

$$\mu_2 = p(1-p) \quad (3.2.6)$$

Thus, the *variance* of the above-mentioned distribution is $p(1-p)$

$$\mu_3 = \mu_3' - 3\mu_2'\mu_1' + 2\mu_1'^3 = p - 3pp + 2p^3 \quad \text{[using Eqns. (3.2.1)–(3.2.3)]}$$

$$= p(1-3p+2p^2) = p(2p^2 - 2p - p + 1)$$

$$\mu_3 = p(1-2p)(1-p) \quad (3.2.7)$$

$$\mu_4 = \mu_4' - 4\mu_3'\mu_1' + 6\mu_2'\mu_1'^2 - 3\mu_1'^4 = p - 4pp + 6pp^2 - 3p^4$$

$$= p\left(1-4p+6p^2-3p^3\right) \quad \text{[using Eqns. (3.2.1)–(3.2.4)]}$$

$$= p(1-p-3p+6p^2-3p^3) = p\{1-p-3p(1-2p+p^2)\}$$

$$= p\{1-p-3p(1-p)^2\}$$

$$\mu_4 = p(1-p)\{1-3p(1-p)\} \quad (3.2.8)$$

c. **Skewness and Kurtosis of Bernoulli Distribution:** The expressions of skewness (β_1, γ_1) and kurtosis (β_2, γ_2) four coefficients based upon the first four central moments are given below (Figure 3.3):

$$\beta_1 = \frac{\mu_3^2}{\mu_2^3} = \frac{[p(1-2p)(1-p)]^2}{[p(1-p)]^3} \quad \text{[using Eqns. (3.2.6) and (3.2.7)]}$$

$$\beta_1 = \frac{(1-2p)^2}{p(1-p)}$$

$$\gamma_1 = \frac{\mu_3}{\mu_2^{3/2}} = \frac{1-2p}{\sqrt{p(1-p)}} \tag{3.2.9}$$

$$\beta_2 = \frac{\mu_4}{\mu_2^2} = \frac{p(1-p)\{1-3p(1-p)\}}{[p(1-p)]^2} = \frac{\{1-3p(1-p)\}}{p(1-p)}$$

[using Eqns. (3.2.6) and (3.2.8)]

$$\beta_2 = \frac{1}{p(1-p)} - 3$$

$$\gamma_2 = \beta_2 - 3 = \frac{1}{p(1-p)} - 6 = \frac{1-6p(1-p)}{p(1-p)} \tag{3.2.10}$$

For $0 < p < \frac{1}{2}$ (i.e.,$\gamma_1 > 0$), the distribution is positively skewed; for $p = \frac{1}{2}$,$\gamma_1 = 0$, this implies that distribution is symmetric when $p = \frac{1}{2}$. Similarly, the distribution is negatively skewed when $\frac{1}{2} \le p < 1$. Let $p_0 = \frac{1}{2} - \frac{1}{\sqrt{2}} = 0.2113$ and $p_1 = \frac{1}{2} - \frac{1}{\sqrt{12}} = 0.7887$ be two values of the parameter p. For $p \in \{p_0, p_1\}$, the distribution is mesokurtic, for $p \in (0, p_0) \cup (p_1, 1)$, the distribution is leptokurtic and for $p \in (p_0, p_1)$, the distribution is platykurtic.

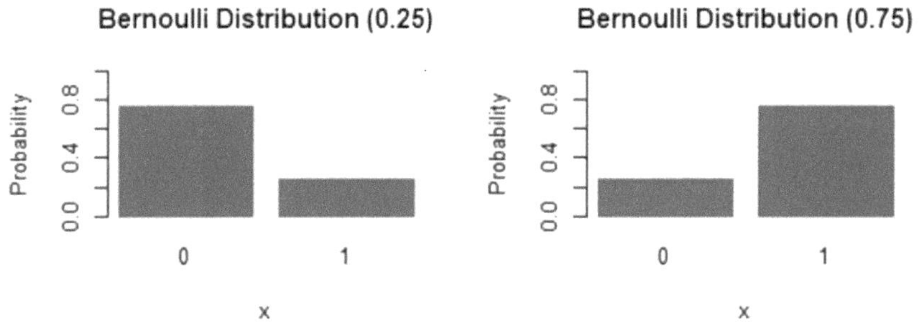

FIGURE 3.3 Shape of Bernoulli distribution.

d. Moment Generating Function (MGF) of Bernoulli Distribution: Each probability distribution has a unique MGF. The expression for MGF ($M_X(t)$) is obtained as below:

$$M_X(t) = E(e^{tX}) = \sum_{x=0}^{1} e^{tx} P(X = x) = \sum_{x=0}^{1} e^{tx} \mathrm{p}^x (1\text{-}\mathrm{p})^{1-x}.$$

$$M_X(t) = (1 - p) + pe^t. \tag{3.2.11}$$

In this method, we find all four moments $\mu_1', \mu_2', \mu_3', \mu_4'$ using $\mu_r' = \frac{\partial^r}{\partial t^r} M_X(t)\Big| t = 0.$

Substitute $r = 1, 2, 3, 4$ and solve them; we can get all four moments $\mu_1', \mu_2', \mu_3', \mu_4'$

$$\mu_1' = \frac{\partial}{\partial t} M_X(t)\Big| t = 0$$

$$\frac{\partial}{\partial t} M_X(t) = \frac{\partial}{\partial t}(1 - p + pe^t) = pe^t \quad \text{[using Eqns. (3.2.11)]} \tag{3.2.12}$$

$$\mu_1' = \left[\frac{\partial}{\partial t} M_X(t)\Big| t = 0 \right] = p$$

$$\mu_2' = \frac{\partial^2}{\partial t^2} M_X(t)\Big| t = 0$$

$$\frac{\partial^2}{\partial t^2} M_X(t) = \frac{\partial}{\partial t}\left[\frac{\partial}{\partial t} M_X(t) \right] = \frac{\partial}{\partial t}(pe^t) = pe^t$$

$$\text{[using Eqns. (3.2.12)]} \tag{3.2.13}$$

$$\mu_2' = \left[\frac{\partial^2}{\partial t^2} M_X(t)\Big| t = 0 \right] = p$$

$$\mu_3' = \frac{\partial^3}{\partial t^3} M_X(t)\Big| t = 0$$

$$\frac{\partial^3}{\partial t^3} M_X(t) = \frac{\partial}{\partial t}\left[\frac{\partial^2}{\partial t^2} M_X(t) \right] = \frac{\partial}{\partial t}\left(pe^t \right) = pe^t$$

$$\text{[using Eqns. (3.2.13)]} \tag{3.2.14}$$

$$\mu_3' = \left[\frac{\partial^3}{\partial t^3} M_X(t)\Big| t = 0 \right] = p$$

$$\mu_4' = \frac{\partial^4}{\partial t^4} M_X(t)\Big| t = 0$$

$$\frac{\partial^4}{\partial t^4}M_X(t) = \frac{\partial}{\partial t}\left[\frac{\partial^3}{\partial t^3}M_X(t)\right] = \frac{\partial}{\partial t}\left(pe^t\right) = pe^t$$

[using Eqns. (3.2.14)] (3.2.15)

$$\mu_4' = \left[\frac{\partial^4}{\partial t^4}M_X(t)\bigg| t=0\right] = p$$

- **Note:** The expression PGF can be derived using the same MGF method. The formula of PGF is $P_X(t) = \sum_x t^x P(X=x)$.

e. **Characteristic Function (CF) of Bernoulli Distribution:** The CF is helpful to know the behaviors of the probability model. The expression for CF ($\phi_X(t)$) is obtained as below:

$$\phi_X(t) = E(e^{itX}) = \sum_{x=0}^{1} e^{itx}P(X=x) = \sum_{x=0}^{1} e^{itx}\text{p}^x(1\text{-p})^{1-x} = \sum_{x=0}^{1}(pe^{it})^x(1\text{-p})^{1-x}$$

$$\phi_X(t) = (1-p)+pe^{it}.$$

- **Note:** We can also generate moments using CF.

f. **Cumulants Generating Function of Bernoulli Distribution:** A CGF takes the moment of a probability density function (PDF) and produces the cumulant. A cumulant of a probability distribution is a sequence of numbers that defines the distribution in a helpful, compact way.
The expression for CGF ($K_X(t)$) is obtained as below:

$$K_X(t) = \log(M_X(t))$$

$$K_X(t) = \log(1-p+pe^t)\,[\text{using Eqns. (3.2.11)}] \quad (3.2.16)$$

We can get central moments using cumulants generating function:

$$\kappa_1 = \text{mean}; \quad \kappa_2 = \text{variance}; \quad \kappa_3 = \mu_3; \quad \kappa_4 = -3\kappa_2^2 + \mu_4.$$

In this method, we find all four κ_1, κ_2, κ_3, κ_4 using $\kappa_r = \frac{\partial^r}{\partial t^r}K_X(t)\Big| t=0.$
Substitute $r = 1, 2, 3, 4$ and solve them; we can get κ_1, κ_2, κ_3, κ_4

$$\kappa_1 = \frac{\partial}{\partial t}K_X(t)\bigg| t=0$$

$$\frac{\partial}{\partial t}K_X(t) = \frac{\partial}{\partial t}\log(1-p+pe^t) = \frac{pe^t}{(1-p+pe^t)}$$

[using Eqns. (3.2.16)] (3.2.17)

$$\kappa_1 = \left[\frac{\partial}{\partial t}K_X(t)\bigg| t=0\right] = p.$$

$$\kappa_2 = \frac{\partial^2}{\partial t^2}K_X(t)\bigg| t=0$$

$$\frac{\partial^2}{\partial t^2}K_X(t) = \frac{\partial}{\partial t}\left[\frac{\partial}{\partial t}K_X(t)\right] = \frac{\partial}{\partial t}\left\{\frac{pe^t}{(1-p+pe^t)}\right\} \text{ [using Eqns. (3.2.17)]}$$

$$\frac{\partial^2}{\partial t^2}K_X(t) = \frac{p\left\{(1-p+pe^t)e^t - pe^te^t\right\}}{(1-p+pe^t)^2} = \frac{(1-p)pe^t}{(1-p+pe^t)^2} \qquad (3.2.18)$$

$$\kappa_2 = \left[\frac{\partial^2}{\partial t^2}M_X(t)\bigg| t=0\right] = p(1-p).$$

$$\kappa_3 = \frac{\partial^3}{\partial t^3}K_X(t)\bigg| t=0$$

$$\frac{\partial^3}{\partial t^3}K_X(t) = \frac{\partial}{\partial t}\left[\frac{\partial^2}{\partial t^2}K_X(t)\right] = \frac{\partial}{\partial t}\left\{\frac{(1-p)pe^t}{(1-p+pe^t)^2}\right\} \text{ [using Eqns. (3.2.18)]}$$

$$= \frac{\partial}{\partial t}\left\{\frac{(1-p)pe^t}{(1-p+pe^t)^2}\right\} = \frac{(1-p)p\left\{(1-p+pe^t)^2e^t - 2(1-p+pe^t)pe^te^t\right\}}{(1-p+pe^t)^4}$$

$$\frac{\partial^3}{\partial t^3}K_X(t) = \frac{(1-p)p(1-p+pe^t)e^t\left\{1-p+pe^t-2pe^t\right\}}{(1-p+pe^t)^4}$$

$$= \frac{(1-p)pe^t\left(1-p-pe^t\right)}{(1-p+pe^t)^3} \qquad (3.2.19)$$

$$\kappa_3 = \left[\frac{\partial^3}{\partial t^3}K_X(t)\bigg| t=0\right] = (1-p)p(1-2p).$$

$$\kappa_4 = \frac{\partial^4}{\partial t^4}K_X(t)\bigg| t=0$$

$$\frac{\partial^4}{\partial t^4}K_X(t) = \frac{\partial}{\partial t}\left[\frac{\partial^3}{\partial t^3}K_X(t)\right] = \frac{\partial}{\partial t}\left\{\frac{(1-p)pe^t\left(1-p-pe^t\right)}{(1-p+pe^t)^3}\right\}$$

[using Eqns. (3.2.19)]

$$= \left[\frac{(1-p)p\left\{(1-p+pe^t)^3\left(e^t(1-p-pe^t)-pe^{2t}\right)-3(1-p+pe^t)^2 pe^t e^t\left(1-p-pe^t\right)\right\}}{(1-p+pe^t)^6} \right]$$

$$= \left[\frac{(1-p)p(1-p+pe^t)^2 e^t\left\{(1-p+pe^t)\left(1-p-2pe^t\right)-3pe^t\left(1-p-pe^t\right)\right\}}{(1-p+pe^t)^6} \right]$$

$$= \left[\frac{(1-p)pe^t\left\{(1-p+pe^t)\left(1-p-2pe^t\right)-3pe^t\left(1-p-pe^t\right)\right\}}{(1-p+pe^t)^4} \right]$$

$$= \frac{(1-p)pe^t}{(1-p+pe^t)^4}\begin{bmatrix} 1-p-2pe^t-p+p^2+2p^2e^t+pe^t \\ -p^2e^t-2p^2e^{2t}-3pe^t+3p^2e^t+3p^2e^{2t} \end{bmatrix}$$

$$= \frac{(1-p)pe^t}{(1-p+pe^t)^4}\left[1-2p+p^2-4pe^t+4p^2e^t+p^2e^{2t}\right]$$

$$\kappa_4 = \left[\frac{\partial^4}{\partial t^4} K_X(t) \bigg| t=0 \right] = (1-p)p\left[1-6p(1-p)\right].$$

g. Recurrence Relation for Moments of Bernoulli Distribution:

$$\mu_{r+1} = p(1-p)\left(r\mu_{r-1} + \frac{d}{dp}\mu_r \right)$$

- **Proof:** The derivation of recurrence relation for the moments of Bernoulli distribution is given below:

$$\mu_r = \sum_x \{x-p\}^r P(X=x) = \sum_{x=0}^{1} \{x-p\}^r \mathrm{p}^x(1\text{-p})^{1-x}$$

Differentiating with respect to p, we get:

$$\frac{d}{dp}\mu_r = \frac{d}{dp}\sum_{x=0}^{1}\{x-p\}^r \mathrm{p}^x(1\text{-p})^{1-x} = \sum_{x=0}^{1}\frac{d}{dp}\{x-p\}^r \mathrm{p}^x(1\text{-p})^{1-x}$$

$$= \sum_{x=0}^{1}\begin{bmatrix} r\{x-p\}^{r-1}(\text{-}1)\mathrm{p}^x(1\text{-p})^{1-x} + \{x-p\}^r \\ \left\{x\mathrm{p}^{x-1}(1\text{-p})^{1-x} + (1-x)\mathrm{p}^x(1\text{-p})^{1-x-1}(-1)\right\} \end{bmatrix}$$

$$= \sum_{x=0}^{1}\left[-r\{x-p\}^{r-1}\mathrm{p}^x(1\text{-p})^{1-x} + \{x-p\}^r \mathrm{p}^x(1\text{-p})^{1-x}\left\{\frac{x-xp-p+xp}{p(1-p)}\right\} \right]$$

$$= \sum_{x=0}^{1}\left[-r\{x-p\}^{r-1}\,\mathrm{p}^{x}(1\text{-p})^{1-x} + \{x-p\}^{r+1}\,\mathrm{p}^{x}(1\text{-p})^{1-x}\left\{\frac{1}{p(1-p)}\right\}\right]$$

$$= -r\sum_{x=0}^{1}\{x-p\}^{r-1}\,\mathrm{p}^{x}(1\text{-p})^{1-x} + \sum_{x=0}^{1}\{x-p\}^{r+1}\,\mathrm{p}^{x}(1\text{-p})^{1-x}\left\{\frac{1}{p(1-p)}\right\}$$

$$\frac{d\mu_r}{dp} = -r\mu_{r-1} + \mu_{r+1}\frac{1}{p(1-p)}$$

$$\mu_{r+1} = p(1-p)\left[\frac{d\mu_r}{dp} + r\mu_{r-1}\right] \qquad (3.2.20)$$

Substituting r = 1 in Eqn. (3.2.20):

$$\mu_{1+1} = p(1-p)\left[\frac{d\mu_1}{dp} + r\mu_0\right]$$

As we know:

$$\mu_0 = 1 \text{ and } \mu_1 = 1$$

$$\mu_2 = p\,(1-p)$$

Substituting r = 2 in Eqn. (3.2.20):

$$\mu_{2+1} = pq\left[\frac{d\mu_2}{dp} + 2\mu_1\right] = p(1-p)\left[\frac{d}{dp}p(1-p)\right]$$

$$\mu_3 = p\,(1-p)\,[1-2p]$$

Substituting r = 3 in Eqn. (3.2.20):

$$\mu_{3+1} = p(1-p)\left[\frac{d\mu_3}{dp} + 3\mu_2\right]$$

$$\mu_4 = p(1-p)\left[\frac{d}{dp}\{p(1-p)(1-2p)\} + 3p(1-p)\right]$$

$$= p(1-p)\left[\frac{d}{dp}\{p-2p^2-p^2+2p^3\} + 3p(1-p)\right]$$

$$= p(1-p)\left[\{1-6p+6p^2\} + 3p(1-p)\right] = p(1-p)(1-3p+3p^2)$$

$$\mu_4 = p(1-p)\{1-3p(1-p)\}$$

3.3 BINOMIAL DISTRIBUTION

It is a generalization of Bernoulli's distribution. If the Bernoulli experiment performs n times, the distribution of the probability of the number of success x occurs in n number of trials is called **Binomial distribution.** It contains a finite number of values (0, 1, 2, …, n) with a probability of success (p) and failure probability (q) with two conditions i) $0 \le p \le 1$ and ii) $p + q = 1$. For example, among 25 HIV tests with two consequences: positive or negative, we can find the probability of five HIV tested positive. This distribution gives the probability of x number of success in n number of trials. In Binomial distribution, the PMF of a random variable X is:

$$P(X = x) = \begin{cases} \dfrac{n!}{x!(n-x)!} p^x q^{n-x} & x=0,1,\ldots,n \\ 0 & \text{otherwise} \end{cases}$$

Symbolically: X~B (n, p).
Here n and p are two parameters of the distribution.

➢ **Sample Generation from Binomial Distribution** (Figure 3.4)

R Code:

$x = rbinom(n, k, p)$
where
$n =$ *sample size*
$k =$ *number of trials*
$p =$ *Probability of success*
Example
$x = rbinom(10,20,.5)$

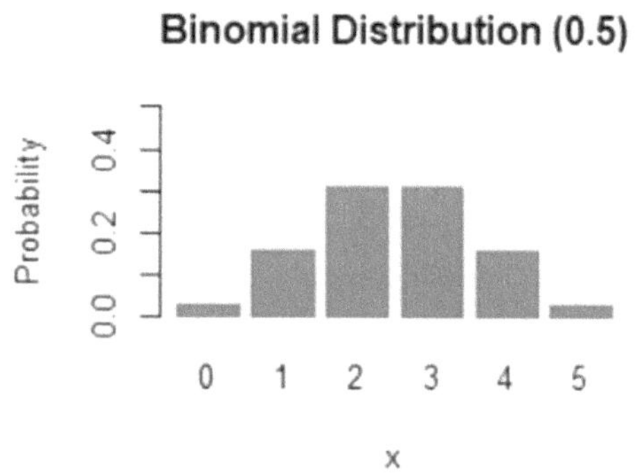

FIGURE 3.4 Binomial distribution.

➢ **Output:**

```
Console  Terminal  Jobs
> x=rbinom(10,20, .5)
> x
 [1] 14 10  9 10 11  6 11 10  6  9
>
```

3.3.1 PROPERTIES OF BINOMIAL DISTRIBUTION

a. Moments of Binomial Distribution: The first four raw moments μ_1', μ_2', μ_3', μ_4' have been derived as below:

$$\mu_1' = E(X) = \sum_{x=0}^{n} xP(X = x) = \sum_{x=0}^{n} x\frac{n!}{x!(n-x)!}\mathrm{p}^x(1\text{-p})^{n-x}$$

$$= \sum_{x=0}^{n} \frac{n!}{(x-1)!(n-x)!}\mathrm{p}^x(1\text{-p})^{n-x}$$

$$= np\sum_{x=1}^{n} \frac{(n-1)!}{(x-1)!(n-x)!}\mathrm{p}^{x-1}(1\text{-p})^{n-x} = np(p+q)^{n-1} = np(1)^{n-1}$$

[using Binomial theorem]

$$\mu_1' = np \tag{3.3.1}$$

Thus, np is the mean of the distribution.

$$\mu_2' = E(X^2) = \sum_{x=0}^{n} x^2P(X = x) = \sum_{x=0}^{n} x^2\frac{n!}{x!(n-x)!}\mathrm{p}^x(1\text{-p})^{n-x}$$

$$= \sum_{x=0}^{n}\left[x(x-1)+x\right]\frac{n!}{x!(n-x)!}\mathrm{p}^x(1\text{-p})^{n-x}$$

$$= \sum_{x=0}^{n} x(x-1)\frac{n!}{(x)!(n-x)!}\mathrm{p}^x(1\text{-p})^{n-x} + \sum_{x=0}^{n} x\frac{n!}{(x)!(n-x)!}\mathrm{p}^x(1\text{-p})^{n-x}$$

$$= n(n-1)p^2\sum_{x=2}^{n} \frac{n-2!}{(x-2)!(n-x)!}\mathrm{p}^{x-2}(1\text{-p})^{n-x} + np \quad \text{[using Eqn. (3.3.1)]}$$

$$= n(n-1)p^2(p+q)^{n-2} + np \quad \text{[using Binomial theorem]}$$

$$\mu_2' = n(n-1)p^2 + np \tag{3.3.2}$$

$$\mu_3' = E(X^3) = \sum_{x=0}^{n} x^3P(X = x) = \sum_{x=0}^{n} x^3\frac{n!}{x!(n-x)!}\mathrm{p}^x(1\text{-p})^{n-x}$$

$$= \sum_{x=0}^{n}\left[x(x-1)(x-2)+3x(x-1)+x\right]\frac{n!}{x!(n-x)!}\mathrm{p}^x(1\text{-p})^{n-x}$$

$$= \left[\begin{array}{l}\sum_{x=0}^{n} x(x-1)(x-2)\frac{n!}{x!(n-x)!}\mathrm{p}^x(1\text{-p})^{n-x} + 3\sum_{x=0}^{n} x(x-1)\frac{n!}{(x)!(n-x)!}\mathrm{p}^x(1\text{-p})^{n-x} \\ +\sum_{x=0}^{n} x\frac{n!}{(x)!(n-x)!}\mathrm{p}^x(1\text{-p})^{n-x}\end{array}\right]$$

$$= \sum_{x=0}^{n} x(x-1)(x-2)\frac{n!}{(x)!(n-x)!}\mathrm{p}^{x}(1\text{-p})^{n-x} + 3n(n-1)p^2 + np$$

[using Eqns. (3.3.1) and (3.3.2)]

$$= n(n-1)(n-2)p^3\sum_{x=3}^{n}\frac{(n-3)!}{(x-3)!(n-x)!}\mathrm{p}^{x-3}(1\text{-p})^{n-x} + 3n(n-1)p^2 + np$$

$$= n(n-1)(n-2)p^3(p+q)^{n-3} + 3n(n-1)p^2 + np \text{ [using Binomial theorem]}$$

$$\mu_3' = n(n-1)(n-2)p^3 + 3n(n-1)p^2 + np \quad (3.3.3)$$

$$\mu_4' = E(X^4) = \sum_{x=0}^{n} x^4 P(X=x) \quad = \sum_{x=0}^{n} x^4\frac{n!}{x!(n-x)!}\mathrm{p}^{x}(1\text{-p})^{n-x}$$

$$= \sum_{x=0}^{n}\left[x(x-1)(x-2)(x-3) + 6x(x-1)(x-2) + 7x(x-1) + x\right]\frac{n!}{x!(n-x)!}\mathrm{p}^{x}(1\text{-p})^{n-x}$$

$$= \begin{bmatrix} \sum_{x=0}^{n} x(x-1)(x-2)(x-3)\frac{n!}{x!(n-x)!}\mathrm{p}^{x}(1\text{-p})^{n-x} + 6\sum_{x=0}^{n} x(x-1)(x-2)\frac{n!}{x!(n-x)!}\mathrm{p}^{x}(1\text{-p})^{n-x} \\ +7\sum_{x=0}^{n} x(x-1)\frac{n!}{(x)!(n-x)!}\mathrm{p}^{x}(1\text{-p})^{n-x} + \sum_{x=0}^{n} x\frac{n!}{(x)!(n-x)!}\mathrm{p}^{x}(1\text{-p})^{n-x} \end{bmatrix}$$

$$= \sum_{x=0}^{n} x(x-1)(x-2)(x-3)\frac{n!}{(x)!(n-x)!}\mathrm{p}^{x}(1\text{-p})^{n-x} + 6n(n-1)$$

$$(n-2)p^3 + 7n(n-1)p^2 + np \quad \text{[using Eqns. (3.3.1)–(3.3.3)]}$$

$$= n(n-1)(n-2)(n-3)p^4\sum_{x=4}^{n}\frac{(n-4)!}{(x-4)!(n-x)!}\mathrm{p}^{x-4}(1\text{-p})^{n-x} + 6n(n-1)$$

$$(n-2)p^3 + 7n(n-1)p^2 + np$$

$$= n(n-1)(n-2)(n-3)p^4(p+q)^{n-4} + 6n(n-1)(n-2)p^3 + 7n(n-1)p^2 + np$$

[using Binomial theorem]

$$\mu_4' = n(n-1)(n-2)(n-3)p^4 + 6n(n-1)(n-2)p^3 + 7n(n-1)p^2 + np \quad (3.3.4)$$

b. **Central Moments Binomial Distribution:** The first four central moments $\mu_1', \mu_2', \mu_3', \mu_4'$ have been derived as below:

$$\mu_1 = 0 \quad (3.3.5)$$

$$\mu_2 = \mu_2' - \mu_1'^2 = n(n-1)p^2 + np - (np)^2 \quad = -np^2 + np = np(1-p)$$

[using Eqns. (3.3.1) and (3.3.2)]

$$\mu_2 = npq \quad (3.3.6)$$

So, the *variance* of the Binomial distribution is *npq*

$$\mu_3 = \mu_3' - 3\mu_2'\mu_1' + 2\mu_1'^3 = \left[n(n-1)(n-2)p^3 + 3n(n-1)p^2 + np\right]$$

$$-3\left[n(n-1)p^2 + np\right]np + 2(np)^3 \qquad \text{[using Eqns. (3.3.1)–(3.3.3)]}$$

$$= np\left[(n-1)(n-2)p^2 + 3(n-1)p + 1 - 3n(n-1)p^2 - 3np + 2n^2p^2\right]$$

$$= np\left[2p^2 - 3p + 1\right]$$

$$\mu_3 = npq\,(q-p) \qquad (3.3.7)$$

$$\mu_4 = \mu_4' - 4\mu_3'\mu_1' + 6\mu_2'\mu_1'^2 - 3\mu_1'^4$$

$$= \begin{bmatrix} n(n-1)(n-2)(n-3)p^4 + 6n(n-1)(n-2)p^3 + 7n(n-1)p^2 + np \\ -4np\left[n(n-1)(n-2)p^3 + 3n(n-1)p^2 + np\right] + 6\left[n(n-1)p^2 + np\right] \\ (np)^2 - 3(np)^4 \end{bmatrix}$$

[using Eqns. (3.3.1)–(3.3.4)]

$$= np\begin{bmatrix} (n-1)(n-2)(n-3)p^3 + 6(n-1)(n-2)p^2 + 7(n-1)p + 1 \\ -4\left[n(n-1)(n-2)p^3 + 3n(n-1)p^2 + np\right] + 6\left[n(n-1)p^2 + np\right] \\ np - 3(np)^3 \end{bmatrix}$$

$$= np\begin{bmatrix} (n^3 - 6n^2 + 11n - 6)p^3 + 6(n^2 - 3n + 2)p^2 + 7(n-1)p + 1 \\ -4np^3(n^2 - 3n + 2) - 12n(n-1)p^2 - 4np + 6np\left[n^2p^2 - np^2 + np\right] - 3n^3p^3 \end{bmatrix}$$

$$= np\begin{bmatrix} n^3p^3 - 6n^2p^3 + 11np^3 - 6p^3 + 6n^2p^2 - 18np^2 + 12p^2 + 7np - 7p + 1 \\ -4n^3p^3 + 12n^2p^3 - 8np^3 - 12n^2p^2 + 12np^2 - 4np + 6n^3p^3 \\ -6n^2p^3 + 6n^2p^2 - 3n^3p^3 \end{bmatrix}$$

$$= np\left[3np^3 - 6p^3 - 6np^2 + 12p^2 + 3np - 7p + 1\right]$$

$$= np\left[(3n-6)p(p^2 - 2p + 1) + q\right]$$

$$\mu_4 = npq\left[3(n-2)pq + 1\right] \qquad (3.3.8)$$

c. **Skewness and Kurtosis of Binomial Distribution:** The expressions of skewness (β_1, γ_1) and kurtosis (β_2, γ_2) four coefficients based upon the first four central moments are given below (Figure 3.5):

$$\beta_1 = \frac{\mu_3^2}{\mu_2^3} = \frac{\left[npq(q-p)\right]^2}{\left[npq\right]^3} = \frac{(q-p)^2}{npq} \qquad \text{[using Eqns. (3.3.6) and (3.3.7)]}$$

$$\gamma_1 = \frac{\mu_3}{\mu_2^{3/2}} = \frac{(q-p)}{\sqrt{npq}} \tag{3.3.9}$$

$$\beta_2 = \frac{\mu_4}{\mu_2^2} = \frac{npq[3(n-2)pq+1]}{[npq]^2} = \frac{[3(n-2)pq+1]}{npq} = 3 + \frac{1-6pq}{npq}$$

[using Eqns. (3.3.6) and (3.3.8)]

$$\gamma_2 = \beta_2 - 3 = \frac{1-6pq}{npq} \tag{3.3.10}$$

For $0 < p < \frac{1}{2}$ (i.e., $\gamma_1 > 0$), the distribution is positively skewed; for $p = \frac{1}{2}$, $\gamma_1 = 0$, this implies that distribution is symmetric when $p = \frac{1}{2}$. Similarly, the distribution is negatively skewed when $\frac{1}{2} \le p < 1$. Let $p_0 = \frac{1}{2} - \frac{1}{\sqrt{2}} = 0.2113$ and $p_1 = \frac{1}{2} - \frac{1}{\sqrt{12}} = 0.7887$ be two values of the parameter p. For $p \in \{p_0, p_1\}$, the distribution is mesokurtic, for $p \in (0, p_0) \cup (p_1, 1)$, the distribution is leptokurtic and for $p \in (p_0, p_1)$, the distribution is platykurtic.

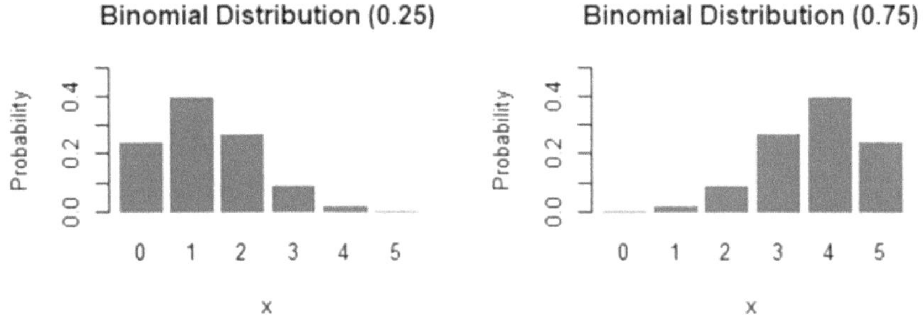

FIGURE 3.5 Shape of the binomial distribution.

d. Moment Generating Function (MGF) of Binomial Distribution: The expression for MGF ($M_x(t)$) is obtained as below:

$$M_X(t) = E(e^{tX}) = \sum_{x=0}^{n} e^{tx} P(X=x) = \sum_{x=0}^{n} e^{tx} \frac{n!}{x!(n-x)!} p^x (1-p)^{n-x}$$

$$= \sum_{x=0}^{n} \frac{n!}{x!(n-x)!} (pe^t)^x (1-p)^{n-x}$$

$$M_X(t) = (1-p+pe^t)^n = \left(q+pe^t\right)^n \quad \text{[using Binomial theorem]} \tag{3.3.11}$$

We can find moments using MGF:

$$\mu_r' = \frac{\partial^r}{\partial t^r} M_X(t) \Big| t=0$$

$$\frac{\partial}{\partial t} M_X(t) = \frac{\partial}{\partial t}(q+pe^t)^n = n(q+pe^t)^{n-1}(pe^t)$$

[using Eqns. (3.3.11)] (3.3.12)

$$\mu_1' = \left[\frac{\partial}{\partial t} M_X(t) \Big| t=0\right] = np$$

$$\mu_2' = \frac{\partial^2}{\partial t^2} M_X(t) \Big| t=0$$

$$\frac{\partial^2}{\partial t^2} M_X(t) = \frac{\partial}{\partial t}\left\{\frac{\partial}{\partial t} M_X(t)\right\} = \frac{\partial}{\partial t}\left[np(q+pe^t)^{n-1}e^t\right]$$

[using Eqn. (3.3.12)] (3.3.12a)

$$\frac{\partial^2}{\partial t^2} M_X(t) = np\left\{(n-1)(q+pe^t)^{n-2}pe^te^t + (q+pe^t)^{n-1}e^t\right\} \tag{3.3.13}$$

$$\mu_2' = \left[\frac{\partial^2}{\partial t^2} M_X(t) \Big| t=0\right] = n(n-1)p^2 + np \tag{3.3.14}$$

$$\mu_3' = \frac{\partial^3}{\partial t^3} M_X(t) \Big| t=0$$

$$\frac{\partial^3}{\partial t^3} M_X(t) = \frac{\partial}{\partial t}\left\{\frac{\partial^2}{\partial t^2} M_X(t)\right\}$$

$$= \frac{\partial}{\partial t}\left[np\left\{(n-1)(q+pe^t)^{n-2}pe^t + (q+pe^t)^{n-1}e^t\right\}\right] \quad \text{[Using Eqn. (3.3.13)]}$$

$$\frac{\partial^3}{\partial t^3} M_X(t) = n(n-1)p^2\frac{\partial}{\partial t}\left\{(q+pe^t)^{n-2}e^t\right\} + np\frac{\partial}{\partial t}\left\{(q+pe^t)^{n-1}e^t\right\}$$

$$\frac{\partial^3}{\partial t^3} M_X(t) = n(n-1)p^2$$

$$\left\{2(q+pe^t)^{n-2}e^{2t} + (n-2)p(q+pe^t)^{n-3}e^te^{2t}\right\} + \frac{\partial^2}{\partial t^2} M_X(t) \tag{3.3.15}$$

$$\mu_3' = \left[\frac{\partial^3}{\partial t^3} M_X(t) \Big| t=0\right] = n(n-1)p^2\left\{2+(n-2)p\right\} + n(n-1)p^2 + np \tag{3.3.16}$$

$$\mu_3' = n(n-1)(n-2)p^3 + 3n(n-1)p^2 + np$$

$$\mu_4' = \frac{\partial^4}{\partial t^4} M_X(t)\Big| t = 0$$

$$\frac{\partial^4}{\partial t^4} M_X(t) = \frac{\partial^4}{\partial t^4}(q + pe^t)^n = \frac{\partial}{\partial t}\left\{\frac{\partial^3}{\partial t^3} M_X(t)\right\}$$

$$\frac{\partial^4}{\partial t^4} M_X(t) = \frac{\partial}{\partial t}$$

$$\left[n(n-1)p^2\left\{2(q+pe^t)^{n-2}e^{2t} + (n-2)p(q+pe^t)^{n-3}e^t e^{2t}\right\} + \frac{\partial^2}{\partial t^2} M_X(t)\right]$$

[using Eqn. (3.3.15)]

$$\frac{\partial^4}{\partial t^4} M_X(t) =$$

$$\left[2n(n-1)p^2 \frac{\partial}{\partial t}(q+pe^t)^{n-2}e^{2t} + n(n-1)(n-2)p^3 \frac{\partial}{\partial t}(q+pe^t)^{n-3}e^{3t} + \frac{\partial}{\partial t}\frac{\partial^2}{\partial t^2} M_X(t)\right]$$

$$\frac{\partial^4}{\partial t^4} M_X(t) =$$

$$\begin{bmatrix} 2n(n-1)p^2\left\{2(q+pe^t)^{n-2}e^{2t} + (n-2)p(q+pe^t)^{n-3}e^t e^{2t}\right\} \\ +n(n-1)(n-2)p^3\left\{3(q+pe^t)^{n-3}e^{3t} + (n-3)p(q+pe^t)^{n-4}e^{3t}\right\} + \frac{\partial^3}{\partial t^3} M_X(t) \end{bmatrix}$$

$$\mu_4' = \left[\frac{\partial^4}{\partial t^4} M_X(t)\Big| t = 0\right]$$

$$= \begin{bmatrix} 2n(n-1)p^2\{2 + (n-2)p\} + n(n-1)(n-2)p^3\{3 + (n-3)p\} \\ +np(n-1)p\{2 + (n-2)p\} + np\{1 + (n-1)p\} \end{bmatrix}$$

[using6 Eqn. (3.3.16)]

$$= \begin{bmatrix} 4n(n-1)p^2 + 2n(n-1)(n-2)p^3 + 3n(n-1)(n-2)p^3 + n(n-1)(n-2) \\ (n-3)p^4 + 2n(n-1)p^2 + n(n-1)(n-2)p^3 + np + n(n-1)p^2 \end{bmatrix}$$

$$\mu_4' = n(n-1)(n-2)(n-3)p^4 + 6n(n-1)(n-2)p^3 + 7n(n-1)p^2 + np$$

- **Note:** The expression PGF can be derived using the same MGF method. The formula of PGF is $P_X(t) = \sum_x t^x P(X = x)$.

e. **Characteristic Function (CF) of Binomial Distribution:** The expression for CF ($\phi_X(t)$) is obtained as below:

$$\phi_X(t) = E(e^{itX}) = \sum_{x=0}^{n} e^{itx} P(X = x) = \sum_{x=1}^{n} e^{itx} \frac{n!}{x!(n-x)!} \mathrm{p}^x \mathrm{q}^{n-x}$$

$$= \sum_{x=1}^{n} \frac{n!}{x!(n-x)!} (pe^{it})^x \mathrm{q}^{n-x}$$

$$\phi_X(t) = \left(q + pe^{it}\right)^n.$$

f. **Cumulant Generating Function (CGF) of Binomial Distribution:** The detailed derivation of the CGF is given below:

$$K_X(t) = \log(M_X(t))$$

$$K_X(t) = n\log(q + pe^t) \quad (3.3.17)$$

We can find central moments using CGF:

➢ **Solution:** $\kappa_r = \frac{\partial^r}{\partial t^r} K_X(t)\Big| t = 0$

Substitute $r = 1, 2, 3, 4$ and solve them; we can get κ_1, κ_2, κ_3, κ_4, and all central moments:

$$\kappa_1 = \text{mean}; \quad \kappa_2 = \text{variance}; \quad \kappa_3 = \mu_3; \quad \kappa_4 = -3\kappa_2^2 + \mu_4.$$

$$\kappa_1 = \text{mean} = \frac{\partial}{\partial t} K_X(t)\Big| t = 0$$

$$\frac{\partial}{\partial t} K_X(t) = \frac{\partial}{\partial t} n\log(q + pe^t) = n\frac{pe^t}{(q + pe^t)} \quad \text{[using Eqn. (3.3.17)]} \quad (3.3.18)$$

$$\kappa_1 = \left[\frac{\partial}{\partial t} K_X(t)\Big| t = 0\right] = \text{mean} = np$$

$$\kappa_2 = \text{variance} = \frac{\partial^2}{\partial t^2} K_X(t)\Big| t = 0$$

$$\frac{\partial^2}{\partial t^2} K_X(t) = \frac{\partial}{\partial t}\left[\frac{\partial}{\partial t} K_X(t)\right] = np\frac{\partial}{\partial t}\left\{\frac{e^t}{(q + pe^t)}\right\}$$

$$= \frac{np\{(q + pe^t)e^t - pe^t e^t\}}{(q + pe^t)^2} = \frac{nqpe^t}{(q + pe^t)^2} \quad \text{[using Eqn. (3.3.18)]}$$

$$\kappa_2 = \left[\frac{\partial^2}{\partial t^2} M_X(t)\bigg| t=0\right] = np(1-p) \qquad (3.3.19)$$

$$\kappa_3 = \mu_3 = \frac{\partial^3}{\partial t^3} K_X(t)\bigg| t=0$$

$$\frac{\partial^3}{\partial t^3} K_X(t) = \frac{\partial}{\partial t}\left[\frac{\partial^2}{\partial t^2} K_X(t)\right] = nqp\frac{\partial}{\partial t}\left\{\frac{e^t}{(q+pe^t)^2}\right\} \quad \text{[using Eqn. (3.3.19)]}$$

$$= \frac{nqp\left\{(q+pe^t)^2 e^t - 2(q+pe^t)pe^{2t}\right\}}{(q+pe^t)^4} = \frac{nqp(q+pe^t)e^t\left\{q+pe^t-2pe^t\right\}}{(q+pe^t)^4}$$

$$\frac{\partial^3}{\partial t^3} K_X(t) = \frac{nqpe^t\left(q-pe^t\right)}{(q+pe^t)^3} \qquad (3.3.20)$$

$$\kappa_3 = \left[\frac{\partial^3}{\partial t^3} K_X(t)\bigg| t=0\right] = nqp(1-2p)$$

$$\kappa_4 = \frac{\partial^4}{\partial t^4} K_X(t)\bigg| t=0$$

$$\frac{\partial^4}{\partial t^4} K_X(t) = \frac{\partial}{\partial t}\left[\frac{\partial^3}{\partial t^3} K_X(t)\right] = nqp\frac{\partial}{\partial t}\left\{\frac{e^t\left(q-pe^t\right)}{(q+pe^t)^3}\right\} \quad \text{[using Eqn. (3.3.20)]}$$

$$=nqp\left[\frac{(q+pe^t)^3\left(e^t(q-pe^t)-pe^{2t}\right)-3(q+pe^t)^2 pe^t e^t\left(q-pe^t\right)}{(q+pe^t)^3}\right]$$

$$= nqp\left[\frac{(q+pe^t)^2 e^t\left\{(q+pe^t)\left(q-2pe^t\right)-3pe^t\left(q-pe^t\right)\right\}}{(q+pe^t)^6}\right]$$

$$= nqp\frac{(q+pe^t)^2 e^t}{(q+pe^t)^6}\left[q^2+pqe^t-2pqe^t-2p^2e^{2t}-3pqe^t+3p^2e^{2t}\right]$$

$$\frac{\partial^4}{\partial t^4} K_X(t) = nqp\frac{(q+pe^t)^2 e^t}{(q+pe^t)^6}\left[q^2-4pqe^t+p^2e^{2t}\right]$$

$$\kappa_4 = \left[\frac{\partial^4}{\partial t^4} K_X(t)\bigg| t=0\right] = nqp\left\{q^2-4pq+p^2\right\} = nqp(1-6pq)$$

$$\mu_4 = \kappa_4 + 3\kappa_2^2 = nqp(1-6pq)+3(npq)^2 = np(1-p)\left\{1+3pq(n-2)\right\}$$

g. Recurrence Relation for the Moments of Binomial Distribution:

$$\mu_{r+1} = pq\left(nr\mu_{r-1} + \frac{d}{dp}\mu_r\right)$$

➢ **Proof:**

$$\mu_r = \sum_x \{x-np\}^r P(X=x) = \sum_{x=0}^{n} \{x-np\}^r \frac{n!}{x!(n-x)!} \mathrm{p}^x (1-p)^{n-x}$$

Differentiating with respect to p, we get:

$$\frac{d}{dp}\mu_r = \frac{d}{dp}\sum_{x=0}^{n} \{x-np\}^r \frac{n!}{x!(n-x)!} \mathrm{p}^x (1-p)^{n-x}$$

$$= \frac{n!}{x!(n-x)!}\sum_{x=0}^{n} \frac{d}{dp}\{x-np\}^r \mathrm{p}^x (1-p)^{n-x}$$

$$= \frac{n!}{x!(n-x)!}\sum_{x=0}^{n}\begin{bmatrix} r\{x-np\}^{r-1}(\text{-n})\mathrm{p}^x(1-p)^{n-x} + \{x-np\}^r \\ \{\mathrm{xp}^{x-1}(1-p)^{n-x} + (n-x)\mathrm{p}^x(1-p)^{n-x-1}(-1)\} \end{bmatrix}$$

$$= \frac{n!}{x!(n-x)!}\sum_{x=0}^{n}\left[-nr\{x-np\}^{r-1}\mathrm{p}^x\mathrm{q}^{n-x} + \{x-np\}^r \mathrm{p}^x\mathrm{q}^{n-x}\left\{\mathrm{x}\frac{1}{p} - (n-x)\frac{1}{q}\right\}\right]$$

$$= \frac{n!}{x!(n-x)!}\sum_{x=0}^{n}\left[-nr\{x-np\}^{r-1}\mathrm{p}^x\mathrm{q}^{n-x} + \{x-np\}^r \mathrm{p}^x\mathrm{q}^{n-x}\left\{\frac{x-np}{pq}\right\}\right]$$

$$= -nr\sum_{x=0}^{n}\{x-np\}^{r-1}\frac{n!}{x!(n-x)!}\mathrm{p}^x\mathrm{q}^{n-x} + \sum_{x=0}^{n}\{x-np\}^{r+1}\frac{n!}{x!(n-x)!}\mathrm{p}^x\mathrm{q}^{n-x}\left\{\frac{1}{pq}\right\}$$

$$\frac{d\mu_r}{dp} = -nr\mu_{r-1} + \mu_{r+1}\frac{1}{pq}$$

$$\mu_{r+1} = pq\left[\frac{d\mu_r}{dp} + nr\mu_{r-1}\right]$$

Substituting r = 1

$$\mu_{1+1} = pq\left[\frac{d\mu_1}{dp} + n\mu_0\right]$$

As we know:

$$\mu_0 = 1 \text{ and } \mu_1 = 0$$

$$\mu_2 = npq$$

Substituting r = 2

$$\mu_{2+1} = pq\left[\frac{d\mu_2}{dp} + 2n\mu_1\right] = pq\left[\frac{d}{dp}np(1-p)\right]$$

$$\mu_3 = npq\ \{1-2p\}$$

Substituting r = 3

$$\mu_{3+1} = pq\left[\frac{d\mu_3}{dp} + 3n\mu_2\right] = pq\left[\frac{d}{dp}\{np(1-p)(q-p)\} + 3n.npq\right]$$

$$= npq\left[\{-2pq + q^2 + p^2 + 2pq - 2pq - 2pq\} + 3npq\right] = npq[-6pq + 1 + 3npq]$$

$$\mu_4 = npq[1 + 3(n-2)pq]$$

3.4 POISSON DISTRIBUTION

If the number of trails is very large and the probability of occurrence of the events is very low, the binomial distribution tends to Poisson distribution. In other words, as $n \to \infty$ and p $\to$ 0 but np = λ (finite, positive real number), the Bin (n, p) approaches Poisson distribution with parameter $\lambda > 0$. It contains a countable number of values (0, 1, 2, 3, ...). For example, the number of deaths due to COVID-19 out of infected patients within one hour. Here we have applied poison distribution because the number of COVID-19 patients is very large, and the probability of the event (death) is very low. In Poisson distribution, the PMF of a random variable X is:

$$P(X = x) = \begin{cases} \dfrac{\lambda^x e^{-\lambda}}{x!} & \text{x=0,1,2,3,...} \\ 0 & \text{otherwise} \end{cases}$$

Symbolically: $X \sim P\ (\lambda)$.

Here λ is the parameter of the distribution and $\lambda > 0$.

➢ **Sample Generation from Poisson Distribution** (Figure 3.6)

R Code:

$x = rpois(n,k)$
where
$n = sample\ size$
$k = parameter\ of\ the\ dist^n$
Example
$x = rpois(10, 4)$

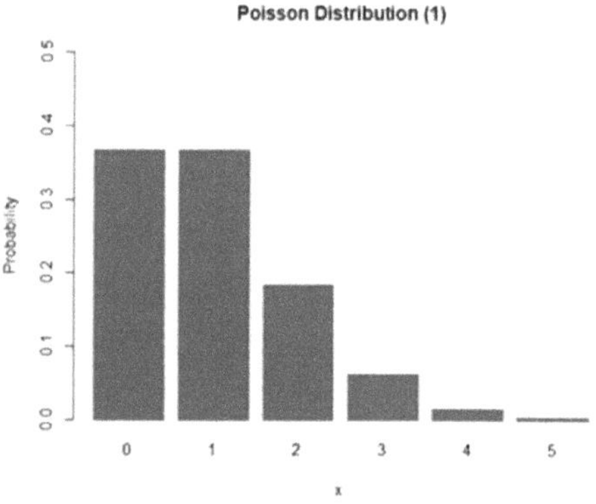

FIGURE 3.6 Poisson distribution.

➢ **Output:**

3.4.1 PROPERTIES OF THE POISSON DISTRIBUTION

a. **Moments of the Poisson Distribution:** The first four raw moments $\mu_1', \mu_2', \mu_3', \mu_4'$ have been derived as below:

$$\mu_1' = E(X) = \sum_{x=0} xP(X = x) = \sum_{x=0} x\frac{\lambda^x e^{-\lambda}}{x!} = \lambda e^{-\lambda}\sum_{x=1}\frac{\lambda^{x-1}}{(x-1)!}$$

Using exponential series, we get:

$$\mu_1' = \lambda \tag{3.4.1}$$

Thus, λ is the mean of the distribution.

$$\mu_2' = E(X^2) = \sum_{x=0} x^2 P(X = x) = \sum_{x=0} x^2\frac{\lambda^x e^{-\lambda}}{x!} = \sum_{x=0}[x(x-1)+x]\frac{\lambda^x e^{-\lambda}}{x!}$$

$$= \sum_{x=0} x(x-1)\frac{\lambda^x e^{-\lambda}}{x!} + \sum_{x=0} x\frac{\lambda^x e^{-\lambda}}{x!} = \sum_{x=2}^{n}\frac{\lambda^x e^{-\lambda}}{x-2!} + \lambda = \lambda^2 e^{-\lambda}\sum_{x=2}^{n}\frac{\lambda^{x-2}}{x-2!} + \lambda$$

Using exponential series and Eqn. (3.4.1), we get:

$$\mu_2' = \lambda^2 + \lambda \tag{3.4.2}$$

$$\mu_3' = E\left(X^3\right) = \sum_{x=0} x^3 P(X = x) = \sum_{x=0} x^3 \frac{\lambda^x e^{-\lambda}}{x!}$$

$$= \sum_{x=0} \left[x(x-1)(x-2) + 3x(x-1) + x\right] \frac{\lambda^x e^{-\lambda}}{x!}$$

$$= \sum_{x=0} x(x-1)(x-2) \frac{\lambda^x e^{-\lambda}}{x!} + \sum_{x=0} 3x(x-1) \frac{\lambda^x e^{-\lambda}}{x!}$$

$$+ \sum_{x=0} x \frac{\lambda^x e^{-\lambda}}{x!} = \lambda^3 e^{-\lambda} \sum_{x=3} \frac{\lambda^{x-3}}{x-3!} + 3\lambda^2 + \lambda$$

[using Eqns. (3.4.1) and (3.4.2)]

Using exponential series, we get:

$$\mu_3' = \lambda^3 + 3\lambda^2 + \lambda \tag{3.4.3}$$

$$\mu_4' = E\left(X^4\right) = \sum_{x=0} x^4 P(X = x) = \sum_{x=0} x^4 \frac{\lambda^x e^{-\lambda}}{x!}$$

$$= \sum_{x=0} \left[x(x-1)(x-2)(x-3) + 6x(x-1)(x-2) + 7x(x-1) + x\right] \frac{\lambda^x e^{-\lambda}}{x!}$$

$$= \sum_{x=0} x(x-1)(x-2)(x-3) \frac{\lambda^x e^{-\lambda}}{x!} + 6\sum_{x=0} x(x-1)(x-2) \frac{\lambda^x e^{-\lambda}}{x!}$$

$$+ 7\sum_{x=0} x(x-1) \frac{\lambda^x e^{-\lambda}}{x!} + \sum_{x=0} x \frac{\lambda^x e^{-\lambda}}{x!}$$

$$= \sum_{x=0} x(x-1)(x-2)(x-3) \frac{\lambda^x e^{-\lambda}}{x!} + 6\lambda^3 + 7\lambda^2 + \lambda$$ [using Eqns. (3.4.1)–(3.4.3)]

$$= \lambda^4 e^{-\lambda} \sum_{x=4} \frac{\lambda^{x-4}}{x-4!} + 6\lambda^3 + 7\lambda^2 + \lambda = \lambda^4 e^{-\lambda} e^{\lambda} + 6\lambda^3 + 7\lambda^2 + \lambda$$

[using exponential series]

$$\mu_4' = \lambda^4 + 6\lambda^3 + 7\lambda^2 + \lambda \tag{3.4.4}$$

b. **Central Moments Poisson Distribution:** The first four central moments μ_1, μ_2, μ_3, μ_4 have been derived as below:
Since, the first central moment about mean is zero, i.e.:

$$\mu_1 = 0 \tag{3.4.5}$$

$$\mu_2 = \mu_2' - \mu_1'^2 = \lambda^2 + \lambda - (\lambda)^2$$ [using Eqns. (3.4.1) and (3.4.2)] (3.4.6)

$$\mu_2 = \lambda$$

So, *variance* of the Poisson distribution is λ.

$$\mu_3 = \mu_3' - 3\mu_2'\mu_1' + 2\mu_1'^3 = \left[\lambda^3 + 3\lambda^2 + \lambda\right] - 3\left[\lambda^2 + \lambda\right]\lambda + 2\lambda^3$$

[using Eqns. (3.4.1)–(3.4.3)]

$$\mu_3 = \lambda \tag{3.4.7}$$

$$\mu_4 = \mu_4' - 4\mu_3'\mu_1' + 6\mu_2'\mu_1'^2 - 3\mu_1'^4$$

$$= \lambda^4 + 6\lambda^3 + 7\lambda^2 + \lambda - 4\lambda\left[\lambda^3 + 3\lambda^2 + \lambda\right] + 6\left[\lambda^2 + \lambda\right]\lambda^2 - 3\lambda^4$$

[using Eqns. (3.4.1)–(3.4.4)]

$$\mu_4 = 3\lambda^2 + \lambda \tag{3.4.8}$$

c. **Skewness and Kurtosis of Poisson Distribution:** The expressions of skewness (β_1, γ_1) and kurtosis (β_2, γ_2) four coefficients based upon the first four central moments are given below (Figure 3.7):

$$\beta_1 = \frac{\mu_3^2}{\mu_2^3} = \frac{[\lambda]^2}{[\lambda]^3} = \frac{1}{\lambda} \quad \text{[using Eqns. (3.4.6) and (3.4.7)]}$$

$$\gamma_1 = \sqrt{\beta_1} = \frac{1}{\sqrt{\lambda}} \tag{3.4.9}$$

$$\beta_2 = \frac{\mu_4}{\mu_2^2} = \frac{3\lambda^2 + \lambda}{[\lambda]^2} = 3 + \frac{1}{\lambda} \quad \text{[using Eqns. (3.4.6) and (3.4.8)]}$$

$$\gamma_2 = \beta_2 - 3 = \frac{1}{\lambda} \tag{3.4.10}$$

This distribution is positive skewed and leptokurtic for all values of $\lambda > 0$. Note that, as $\lambda \rightarrow$, the distribution approaches to symmetric and mesokurtic.

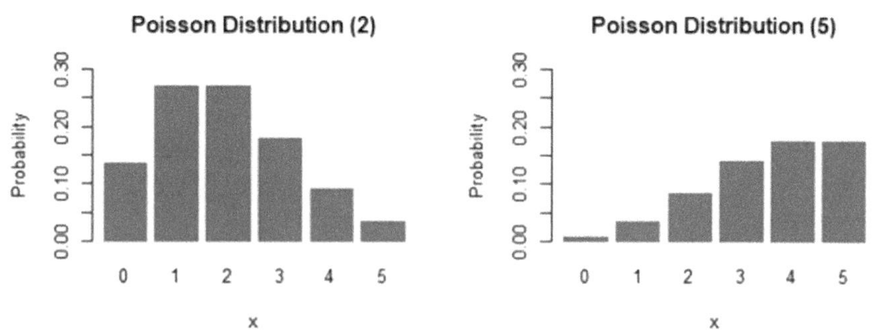

FIGURE 3.7 Shape of the Poisson distribution.

d. Moment Generating Function (MGF) of Poisson Distribution: The expression for MGF ($M_X(t)$ is obtained as below:

$$M_X(t) = E(e^{tX}) = \sum_{x=0} e^{tx} P(X = x) = \sum_{x=0} e^{tx} \frac{\lambda^x e^{-\lambda}}{x!} = \sum_{x=0} \frac{(e^t \lambda)^x e^{-\lambda}}{x!} = e^{-\lambda} \sum_{x=0} \frac{(e^t \lambda)^x}{x!}$$

$$M_X(t) = e^{\lambda(e^t - 1)} \quad \text{[using exponential series]} \quad (3.4.11)$$

We can find moments using the moment generating function:

$$\mu_r' = \frac{\partial^r}{\partial t^r} M_X(t)\Big| t = 0$$

$$\mu_1' = \frac{\partial}{\partial t} M_X(t)\Big| t = 0$$

$$\frac{\partial}{\partial t} M_X(t) = \frac{\partial}{\partial t} e^{\lambda(e^t - 1)} = e^{\lambda(e^t - 1)} (\lambda e^t) \quad \text{[using Eqns. (3.4.11)]} \quad (3.3.12)$$

$$\mu_1' = \left[\frac{\partial}{\partial t} M_X(t)\Big| t = 0 \right] = \lambda$$

$$\mu_2' = \frac{\partial^2}{\partial t^2} M_X(t)\Big| t = 0$$

$$\frac{\partial^2}{\partial t^2} M_X(t) = \frac{\partial}{\partial t} \left\{ \frac{\partial}{\partial t} M_X(t) \right\} = \frac{\partial}{\partial t} \left[\lambda e^t e^{\lambda(e^t - 1)} \right] \quad \text{[using Eqn. (3.4.12)]}$$

$$\frac{\partial^2}{\partial t^2} M_X(t) = \lambda \left\{ e^t e^{\lambda(e^t - 1)} (\lambda e^t) + e^t e^{\lambda(e^t - 1)} \right\} \quad (3.4.13)$$

$$\mu_2' = \left[\frac{\partial^2}{\partial t^2} M_X(t)\Big| t = 0 \right] = \lambda^2 + \lambda \quad (3.4.14)$$

$$\mu_3' = \frac{\partial^3}{\partial t^3} M_X(t)\Big| t = 0$$

$$\frac{\partial^3}{\partial t^3} M_X(t) = \frac{\partial}{\partial t} \left\{ \frac{\partial^2}{\partial t^2} M_X(t) \right\} = \frac{\partial}{\partial t} \left[\lambda \left\{ e^t e^{\lambda(e^t - 1)} (\lambda e^t) + e^t e^{\lambda(e^t - 1)} \right\} \right]$$

[using Eqn. (3.4.13)]

$$\frac{\partial^3}{\partial t^3} M_X(t) = \lambda^2 \frac{\partial}{\partial t} \left\{ e^{2t} e^{\lambda(e^t - 1)} \right\} + \frac{\partial}{\partial t} \left\{ \lambda e^t e^{\lambda(e^t - 1)} \right\} = \lambda^2 \frac{\partial}{\partial t} \left\{ e^{2t} e^{\lambda(e^t - 1)} \right\} + \frac{\partial}{\partial t} \left\{ \frac{\partial}{\partial t} M_X(t) \right\}$$

[using Eqn. (3.4.12)]

$$\frac{\partial^3}{\partial t^3} M_X(t) = \lambda^2 \left\{2e^{\lambda(e^t-1)}e^{2t} + \lambda e^{\lambda(e^t-1)}e^{2t}e^t\right\} + \frac{\partial^2}{\partial t^2} M_X(t) \qquad (3.4.15)$$

$$\mu_3' = \left[\frac{\partial^3}{\partial t^3} M_X(t)\bigg| t=0\right] = \lambda^2\{2+\lambda\} + \lambda^2 + \lambda$$

$$\mu_3' = \lambda^3 + 3\lambda^2 + \lambda \qquad (3.4.16)$$

$$\mu_4' = \frac{\partial^4}{\partial t^4} M_X(t)\bigg| t=0$$

$$\frac{\partial^4}{\partial t^4} M_X(t) = \frac{\partial}{\partial t}\left\{\frac{\partial^3}{\partial t^3} M_X(t)\right\} = \frac{\partial}{\partial t}\left[\lambda^2\left\{2e^{\lambda(e^t-1)}e^{2t} + \lambda e^{\lambda(e^t-1)}e^{2t}e^t\right\} + \frac{\partial^2}{\partial t^2} M_X(t)\right]$$

[using Eqn. (3.4.15)]

$$= \left[2\lambda^2 \frac{\partial}{\partial t} e^{\lambda(e^t-1)}e^{2t} + \lambda^3 \frac{\partial}{\partial t} e^{\lambda(e^t-1)}e^{3t} + \frac{\partial}{\partial t}\frac{\partial^2}{\partial t^2} M_X(t)\right]$$

$$\frac{\partial^4}{\partial t^4} M_X(t = \begin{bmatrix} 2\lambda^2\left\{2e^{\lambda(e^t-1)}e^{2t} + \lambda e^{\lambda(e^t-1)}e^t e^{2t}\right\} \\ +\lambda^3\left\{3e^{\lambda(e^t-1)}e^{3t} + \lambda e^{\lambda(e^t-1)}e^t e^{3t}\right\} + \frac{\partial^3}{\partial t^3} M_X(t) \end{bmatrix}$$

$$\mu_4' = \left[\frac{\partial^4}{\partial t^4} M_X(t)\bigg| t=0\right] = \left[2\lambda^2\{2+\lambda\} + \lambda^3\{3+\lambda\} + \lambda^3 + 3\lambda^2 + \lambda\right]$$

[using Eqn. (3.4.16)]

$$\mu_4' = \lambda^4 + 6\lambda^3 + 7\lambda^2 + \lambda$$

➢ **Note:** The expression PGF can be derived using the same MGF method. The formula of PGF is $P_X(t) = \sum_x t^x P(X=x)$.

e. **Characteristic Function (CF) of Poisson Distribution:** The expression for CF ($\phi_X(t)$) is obtained as below:

$$\phi_X(t) = E\left(e^{itX}\right) = \sum_{x=0} e^{itx} P(X=x) = \sum_{x=0} e^{itx}\frac{\lambda^x e^{-\lambda}}{x!} = \sum_{x=0}\frac{(e^{it}\lambda)^x e^{-\lambda}}{x!}$$

$$= e^{-\lambda}\sum_{x=0}\frac{(e^{it}\lambda)^x}{x!}$$

$$\phi_X(t) = e^{\lambda(e^{it}-1)} \qquad \text{[using exponential series]}$$

f. **Cumulant Generating Function (CGF) of Poisson Distribution:** The detailed derivation of the CDF is given below:

$$K_X(t) = \log(M_X(t))$$

$$K_X(t) = \log e^{\lambda(e^t-1)}$$

$$K_X(t) = \lambda(e^t - 1) \tag{3.4.17}$$

We can find moments using cumulant generating function:

$$\kappa_r = \text{is the coefficient of } \frac{t^r}{r!}.$$

$$\kappa_1 = \text{mean}; \quad \kappa_2 = \text{variance}; \quad \kappa_3 = \mu_3; \quad \kappa_4 = -3\kappa_2 + \mu_4.$$

$$K_X(t) = \lambda\left(1 + \frac{t}{1!} + \frac{t^2}{2!} + \frac{t^3}{3!} + \frac{t^4}{4!} \ldots\ldots - 1\right) = \lambda\left(\frac{t}{1!} + \frac{t^2}{2!} + \frac{t^3}{3!} + \frac{t^4}{4!} \ldots\ldots\right)$$

[using exponential series]

$$\kappa_1 \text{ is the coefficient of } \frac{t}{1!} = \lambda.$$

$$\kappa_2 \text{ is the coefficient of } \frac{t^2}{2!} = \lambda.$$

$$\kappa_3 \text{ is the coefficient of } \frac{t^3}{3!} = \lambda.$$

$$\kappa_4 \text{ is the coefficient of } \frac{t^4}{4!} = \lambda.$$

g. Recurrence Relation for Moments of Poisson Distribution:

$$\mu_{r+1} = \lambda\left[r\mu_{r-1} + \frac{d\mu_r}{d\lambda}\right]$$

➢ **Proof:**

$$\mu_r = \sum_x \{x-\lambda\}^r P(X=x) = \sum_{x=0} \{x-\lambda\}^r \frac{\lambda^x e^{-\lambda}}{x!}$$

Differentiating with respect to λ, we get:

$$\frac{d}{d\lambda}\mu_r = \frac{d}{d\lambda}\sum_{x=1}^{n} \{x-\lambda\}^r \frac{\lambda^x e^{-\lambda}}{x!} = \sum_{x=0} \frac{1}{x!}\frac{d}{d\lambda}\{x-\lambda\}^r \lambda^x e^{-\lambda}$$

$$= \sum_{x=0} \frac{1}{x!}\left[r\{x-\lambda\}^{r-1}(-1)\lambda^x e^{-\lambda} + \{x-\lambda\}^r\left(\lambda^x e^{-\lambda}(-1) + x\lambda^{x-1}e^{-\lambda}\right)\right]$$

$$= \sum_{x=0} \frac{1}{x!}\left[-r\{x-\lambda\}^{r-1}\lambda^x e^{-\lambda} + \{x-\lambda\}^r \lambda^x e^{-\lambda}\left(-1 + \frac{x}{\lambda}\right)\right]$$

$$= \sum_{x=0} \frac{1}{x!}\left[-r\{x-\lambda\}^{r-1}\lambda^x e^{-\lambda} + \{x-\lambda\}^r \lambda^x e^{-\lambda}\left(\frac{x-\lambda}{\lambda}\right)\right]$$

$$= \sum_{x=0} \frac{1}{x!}\left[-r\{x-\lambda\}^{r-1}\lambda^x e^{-\lambda} + \{x-\lambda\}^{r+1} \lambda^x e^{-\lambda}\left(\frac{1}{\lambda}\right)\right]$$

$$= -r\sum_{x=0}\left[\{x-\lambda\}^{r-1}\frac{1}{x!}\lambda^x e^{-\lambda}\right] + \sum_{x=0}\left[\{x-\lambda\}^{r+1}\frac{1}{x!}\lambda^x e^{-\lambda}\left(\frac{1}{\lambda}\right)\right]$$

$$\frac{d\mu_r}{d\lambda} = -r\mu_{r-1} + \left(\frac{1}{\lambda}\right)\mu_{r+1}$$

$$\mu_{r+1} = \lambda\left[r\mu_{r-1} + \frac{d\mu_r}{d\lambda}\right]$$

Substituting r = 1

$$\mu_{1+1} = \lambda\left[\mu_0 + \frac{d\mu_1}{d\lambda}\right]$$

As we know:

$$\mu_0 = E(x - Ex)^0 = E(1) = 1 \text{ and } \mu_1 = 0$$

$$\mu_2 = \lambda\ [1 + 0]$$

$$\mu_2 = \lambda$$

Substituting r = 2

$$\mu_{2+1} = \lambda\left[\frac{d\mu_2}{d\lambda} + 2\mu_1\right] = \lambda\left[\frac{d}{d\lambda}\lambda + 0\right]$$

$$\mu_3 = \lambda$$

Substituting r = 3

$$\mu_{3+1} = \lambda\left[\frac{d\mu_3}{d\lambda} + 3\mu_2\right] = \lambda\left[\frac{d(\lambda)}{d\lambda} + 3\lambda\right] = \lambda[1 + 3\lambda]$$

$$\mu_4 = \lambda + 3\lambda^2$$

3.5 GEOMETRIC DISTRIBUTION

Geometric distribution contains a countable number of values (0, 1, 2, …) with success probability (p) and failure probability (q) with two conditions

i) $0 \le p \le 1$ and ii) $p + q = 1$. This distribution gives the probability of the number of failures before the first success (x). For example, the number of personal loan offers given by the lender to the customers until the first loan is sanctioned. In Geometric distribution, the PMF of a random variable X is:

$$P(X = x) = \begin{cases} pq^x & \text{x=0,1,2...} \\ 0 & \text{otherwise} \end{cases}$$

Symbolically: $X \sim Geo(p)$.

Here p is the parameter of the distribution.

➢ **Sample Generation from Geometric Distribution** (Figure 3.8)

R Code:

$x = rgeom(n, k)$
where
$n = sample\ size$
$k = parameter\ of\ the\ distribution$
Example
$x = rgeom(10{,}0.5)$

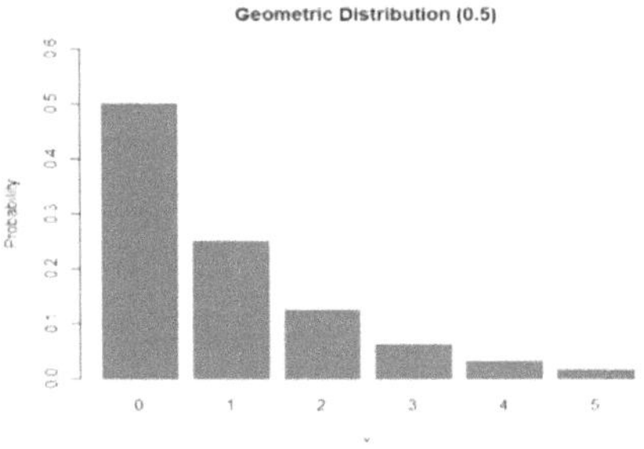

FIGURE 3.8 Geometric distribution.

➢ **Output:**

```
Console  Terminal  Jobs
> x= rgeom(10,0.5)
> x
 [1] 0 2 0 1 1 0 1 0 1 3
>
```

3.5.1 PROPERTIES OF THE GEOMETRIC DISTRIBUTION

a. **Moments of Geometric Distribution:** The first four raw moments μ'_1, μ'_2, μ'_3, μ'_4 have been derived as below:

$$\mu'_1 = E(X) = \sum_x xP(X = x) = \sum_{x=0} xpq^x = p(0q^0 + 1q^1 + 2q^2 + 3q^3.......)$$

$$\mu'_1 = p(q + 2q^2 + 3q^3.......)$$

Multiply by q in the above equation:

$$q\mu_1' = p(q^2 + 2q^3 + 3q^4 ....)$$

Subtracting the above two equations, we have:

$$\mu_1' - q\mu_1' = p(q + 2q^2 + 3q^3 ....) - p(q^2 + 2q^3 + 3q^4 ....)$$

$$\mu_1'(1-q) = p(q + q^2 + q^3 ....)$$

$$\mu_1'(1-q) = p\frac{q}{1-q}$$

$$\mu_1' = \frac{pq}{(1-q)^2}$$

$$\mu_1' = \frac{q}{p} \tag{3.5.1}$$

Thus, $\frac{q}{p}$ is the mean of the distribution.

$$\mu_2' = E(X^2) = \sum_x x^2 P(X = x) =$$

$$\sum_{x=0} x^2 pq^x = p(0^2 q^0 + 1^2 q^1 + 2^2 q^2 + 3^2 q^3 + 4^2 q^4 + 5^2 q^5 ....)$$

$$\mu_2' = p(0 + 1q^1 + 4q^2 + 9q^3 + 16q^4 + 25q^5 ....)$$

Multiply by q in the above equation:

$$q\mu_2' = p(1q^2 + 4q^3 + 9q^4 + 16q^5 + 25q^6 ....)$$

Subtracting the above two equations, we have:

$$\mu_2' - q\mu_2' = p(0 + 1q^1 + 4q^2 + 9q^3 + 16q^4 + 25q^5 ....)$$
$$- p(1q^2 + 4q^3 + 9q^4 + 16q^5 + 25q^6 ....)$$

$$\mu_2'(1-q) = p(q + 3q^2 + 5q^3 + 7q^4 ....)$$

Multiply by q in the above equation:

$$\mu_2' q(1-q) = p(q^2 + 3q^3 + 5q^4 + 7q^5 ....)$$

Subtracting the above two equations, we have:

$$\mu_2'(1-q)^2 = p(q + 2q^2 + 2q^3 + 2q^4 ....) = pq + 2q^2 p(1 + q + q^2 + q^3 ....)$$

$$\mu_2'(1-q)^2 = pq + 2q^2 p\frac{1}{1-q} = \frac{1}{p^2}\left[pq + 2q^2\right] = \frac{1}{p^2}\left[q(p+q) + q^2\right]$$

$$\mu_2' = \frac{q}{p^2}[q+1] \tag{3.5.2}$$

$$\mu_3' = E(X^3) = \sum_{x=0} x^3 pq^x = p(0^3 q^0 + 1^3 q^1 + 2^3 q^2 + 3^3 q^3 + 4^3 q^4 + 5^3 q^5.......)$$

$$\mu_3' = p(0 + 1q^1 + 8q^2 + 27q^3 + 64q^4 + 125q^5.......)$$

Multiply by q in the above equation:

$$q\mu_3' = p(q^2 + 8q^3 + 27q^4 + 64q^5 + 125q^6.......)$$

Subtracting the above two equations, we have:

$$\mu_3' - q\mu_3' = p(0 + 1q^1 + 8q^2 + 27q^3 + 64q^4 + 125q^5.......)$$
$$-p(q^2 + 8q^3 + 27q^4 + 64q^5 + 125q^6.......)$$
$$\mu_3'(1-q) = p(q + 7q^2 + 19q^3 + 37q^4.......)$$

Multiply by q in the above equation:

$$\mu_3' q(1-q) = p(q^2 + 7q^3 + 19q^4 + 37q^5.......)$$

Subtracting the above two equations, we have:

$$\mu_3'(1-q)^2 = p(q + 6q^2 + 12q^3 + 18q^4.......)$$

Multiply by q in the above equation:

$$\mu_3' q(1-q)^2 = p(q^2 + 6q^3 + 12q^4 + 18q^5.......)$$

Subtracting the above two equations, we have:

$$\mu_3'(1-q)^3 = p(q + 6q^2 + 12q^3 + 18q^4.......)$$
$$-p(q^2 + 6q^3 + 12q^4 + 18q^5.......)$$

$$\mu_3'(1-q)^3 = p(q + 5q^2 + 6q^3 + 6q^4.......)$$

$$\mu_3'(1-q)^3 = p\{q + 5q^2 + 6q^3(1 + q + q^2....)\} = p\left\{q + 5q^2 + 6q^3 \frac{1}{1-q}\right\}$$

$$\mu_3' = \frac{q}{p^3}\{p + 5pq + 6q^2\} = \frac{q}{p^3}\{p + 5pq + 5q^2 + q^2\} = \frac{q}{p^3}\left(1 + 4q + q^2\right) \tag{3.5.3}$$

$$\mu_4' = E(X^4) = \sum_{x=0} x^4 pq^x = p(0^4 q^0 + 1^4 q^1 + 2^4 q^2 + 3^4 q^3 + 4^4 q^4 + 5^4 q^5.......)$$

$$\mu_4' = p(0 + 1q^1 + 16q^2 + 81q^3 + 256q^4 + 625q^5.......)$$

Multiply by q in the above equation:

$$q\mu_4' = p(0+1q^2+16q^3+81q^4+256q^5+625q^6.......)$$

Subtracting above two equations, we have:

$$\mu_4' - q\mu_4' = p(0+1q^1+16q^2+81q^3+256q^4+625q^5.......)$$
$$-p(0+1q^2+16q^3+81q^4+256q^5+625q^6.......)$$

$$\mu_4'(1-q) = p(q+15q^2+65q^3+175q^4+369q^5.......)$$

Multiply by q in the above equation:

$$\mu_4' q(1-q) = p(q^2+15q^3+65q^4+175q^5+369q^6.......)$$

Subtracting above two equations, we have:

$$\mu_4'(1-q)^2 = p(q+15q^2+65q^3+175q^4+369q^5.......)$$
$$-p(q^2+15q^3+65q^4+175q^5+369q^6.......)$$

$$\mu_4'(1-q)^2 = p(q+14q^2+50q^3+110q^4+194q^5.......)$$

Multiply by q in the above equation:

$$\mu_4' q(1-q)^2 = p(q^2+14q^3+50q^4+110q^5+194q^6.......)$$

Subtracting above two equations, we have:

$$\mu_4'(1-q)^3 = p(q+14q^2+50q^3+110q^4+194q^5.......)$$
$$-p(q^2+14q^3+50q^4+110q^5+194q^6.......)$$
$$\mu_4'(1-q)^3 = p(q+13q^2+36q^3+60q^4+84q^5.......)$$

Multiply by q in the above equation:

$$\mu_4' q(1-q)^3 = p(q^2+13q^3+36q^4+60q^5+84q^6.......)$$

Subtracting the above two equations, we have:

$$\mu_4'(1-q)^4 = p(q+13q^2+36q^3+60q^4+84q^5.......)$$
$$-p(q^2+13q^3+36q^4+60q^5+84q^6.......)$$

$$\mu_4'(1-q)^4 = p(q+12q^2+13q^3+24q^4+24q^5.......)$$

Multiply by q in the above equation:

$$\mu_4' q(1-q)^4 = p(q^2+12q^3+13q^4+24q^5+24q^6.......)$$

Subtracting the above two equations, we have:

$$\mu_4'(1-q)^5 = p(q+12q^2+13q^3+24q^4+24q^5.......)$$
$$-p(q^2+12q^3+13q^4+24q^5+24q^6.......)$$

$$\mu_4'(1-q)^5 = p(q+11q^2+q^3+11q^4)$$

$$\mu_4' = \frac{q}{p^4}(1+11q+11q^2+q^3) \quad (3.5.4)$$

b. **Central Moments of Geometric Distribution:** The first four central moments μ_1, μ_2, μ_3, μ_4 have been derived as below:

$$\mu_1 = 0 \quad (3.4.5)$$

$$\mu_2 = \mu_2' - \mu_1'^2 = \frac{q}{p^2}(q+1) - \left(\frac{q}{p}\right)^2 \quad \text{[using Eqns. (3.5.1) and (3.5.2)]}$$

$$\mu_2 = \frac{q}{p^2} \quad (3.5.6)$$

So, *variance* of the geometric distribution is $\frac{q}{p^2}$.

$$\mu_3 = \mu_3' - 3\mu_2'\mu_1' + 2\mu_1'^3 = \frac{q}{p^3}\{1+4q+q^2\} - 3\left(\frac{q}{p^2}(q+1)\right)\left(\frac{q}{p}\right) + 2\left(\frac{q}{p}\right)^3$$

[using Eqns. (3.5.1)–(3.5.3)]

$$\mu_3 = \frac{q}{p^3}(1+q) \quad (3.5.7)$$

$$\mu_4 = \mu_4' - 4\mu_3'\mu_1' + 6\mu_2'\mu_1'^2 - 3\mu_1'^4$$

$$= \frac{q}{p^4}\left(1+11q+11q^2+q^3\right) - 4\frac{q}{p^3}\left(1+4q+q^2\right)\left(\frac{q}{p}\right) + 6\frac{q}{p^2}(q+1)\left(\frac{q}{p}\right)^2 - 3\left(\frac{q}{p}\right)^4$$

[using Eqns. (3.5.1)–(3.5.4)]

$$= \frac{q}{p^4}\left[1+11q+11q^2+q^3-4\left(q+4q^2+q^3\right)+6\left(q^3+q^2\right)-3q^3\right]$$

$$\mu_4 = \frac{q}{p^4}\left[1+7q+7q^2\right] \quad (3.5.8)$$

c. **Skewness and Kurtosis of Geometric Distribution:** The expressions of skewness (β_1, γ_1) and kurtosis (β_2, γ_2) four coefficients based upon the first four central moments are given below (Figure 3.9):

$$\beta_1 = \frac{\mu_3^2}{\mu_2^3} = \frac{\left[\frac{q(1+q)}{p^3}\right]^2}{\left[\frac{q}{p^2}\right]^3} = \frac{(1+q)^2}{q}$$

$$\gamma_1 = \frac{\mu_3}{\mu_2^{3/2}} = \frac{(1+q)}{\sqrt{q}} \tag{3.5.9}$$

$$\beta_2 = \frac{\mu_4}{\mu_2^2} = \frac{\frac{q}{p^4}\left[1+7q+q^2\right]}{\left[\frac{q}{p^2}\right]^2} = \frac{\left[1+7q+q^2\right]}{q} = 7 + \frac{\left[1+q^2\right]}{q} = 7 + \frac{\left[1+(1-p)^2\right]}{q}$$

$$= 7 + \frac{\left[1+1+p^2-2p\right]}{q} = 7 + \frac{\left[2q+p^2\right]}{q}$$

$$\beta_2 = 9 + \frac{p^2}{q}$$

$$\gamma_2 = \beta_2 - 3 = 6 + \frac{p^2}{q} \tag{3.5.10}$$

This distribution is positively skewed and leptokurtic.

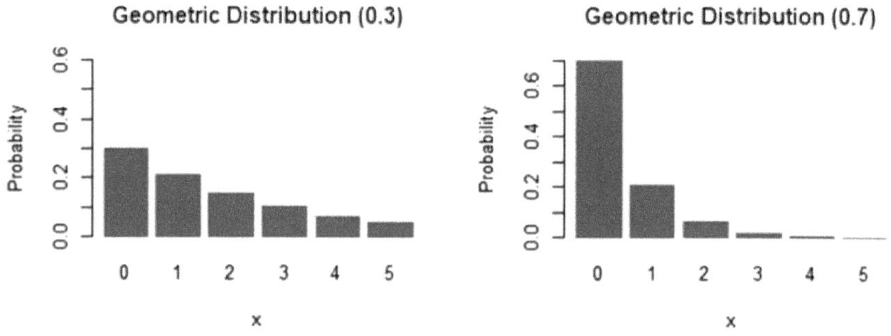

FIGURE 3.9 Shape of the geometric distribution.

The expression for MGF ($M_x(t)$) is obtained as below:

$$M_X(t) = E(e^{tX}) = \sum_{x=0} e^{tx} P(X = x) = \sum_{x=0} e^{tx} pq^x = p\sum_{x=0} (e^t q)^x$$

$$= p\left[(e^t q)^0 + (e^t q)^1 + (e^t q)^2 + ...\right] = p\frac{1}{1-e^t q} \text{ [using sum of geometric series]}$$

$$M_X(t) = p\left(1-qe^t\right)^{-1}, \text{t} < -\ln(\text{q}) \tag{3.5.11}$$

We can get moments using the moment generating function:

$$\mu_r' = \frac{\partial^r}{\partial t^r} M_X(t)\Big| t=0$$

$$\mu_1' = \frac{\partial}{\partial t} M_X(t)\Big| t=0$$

$$\frac{\partial}{\partial t} M_X(t) = p\frac{\partial}{\partial t}(1-qe^t)^{-1} = p(-1)(1-qe^t)^{-1-1}\left(-qe^t\right) = pqe^t(1-qe^t)^{-1-1}$$

[using Eqn. (3.5.11)]

$$\frac{\partial}{\partial t} M_X(t) = pqe^t(1-qe^t)^{-2} \tag{3.5.12}$$

$$\mu_1' = \left[\frac{\partial}{\partial t} M_X(t)\Big| t=0\right] = \frac{q}{p} \tag{3.5.13}$$

$$\mu_2' = \frac{\partial^2}{\partial t^2} M_X(t)\Big| t=0$$

$$\frac{\partial^2}{\partial t^2} M_X(t) = \frac{\partial}{\partial t}\left\{\frac{\partial}{\partial t} p(1-qe^t)^{-1}\right\} = \frac{\partial}{\partial t}\left[pqe^t(1-qe^t)^{-2}\right] = pq\frac{\partial}{\partial t}\left[e^t(1-qe^t)^{-2}\right]$$

[using Eqn. (3.5.11)]

$$= pq\left[(1-qe^t)^{-2}e^t + (-2)(1-qe^t)^{-2-1}(-qe^t)e^t\right]$$

$$\frac{\partial^2}{\partial t^2} M_X(t) = pq\left[(1-qe^t)^{-2}e^t + 2qe^{2t}(1-qe^t)^{-3}\right] \tag{3.5.14}$$

$$\mu_2' = \left[\frac{\partial^2}{\partial t^2} M_X(t)\Big| t=0\right] = pq\left[\frac{1}{p^2} + \frac{2q}{p^3}\right] = \frac{q}{p^2}(1+q) \tag{3.5.15}$$

$$\mu_3' = \frac{\partial^3}{\partial t^3} M_X(t)\Big| t=0$$

$$\frac{\partial^3}{\partial t^3} M_X(t) = \frac{\partial^3}{\partial t^3} p(1-qe^t)^{-1} = \frac{\partial}{\partial t}\left[\frac{\partial^2}{\partial t^2} p(1-qe^t)^{-1}\right]$$

$$= pq\frac{\partial}{\partial t}\left[\left\{(1-qe^t)^{-2}e^t + 2qe^{2t}(1-qe^t)^{-3}\right\}\right]$$

$$= pq\left[\frac{\partial}{\partial t}(1-qe^t)^{-2}e^t + 2q\frac{\partial}{\partial t}e^{2t}(1-qe^t)^{-3}\right] \quad \text{[using Eqn. (3.5.5)]}$$

$$=pq\left[(1-qe^t)^{-2}e^t + 2(1-qe^t)^{-3}e^t\left(qe^t\right) + 4qe^{2t}(1-qe^t)^{-3} + 6qe^{2t}(1-qe^t)^{-4}\left(qe^t\right)\right]$$

$$\frac{\partial^3}{\partial t^3}M_X(t) = pq\left[(1-qe^t)^{-2}e^t + 6qe^{2t}(1-qe^t)^{-3} + 6q^2e^{3t}(1-qe^t)^{-4}\right] \quad (3.5.16)$$

$$\mu_3' = \left[\frac{\partial^3}{\partial t^3}M_X(t)\bigg| t=0\right] = pq\left[\frac{1}{p^2}+\frac{6q}{p^3}+\frac{6q^2}{p^4}\right]$$

$$= \frac{q}{p^3}\left[p^2+6qp+6q^2\right] = \frac{q}{p^3}\left[p^2+6q\right] \quad \text{[using Eqn. (3.5.6)]}$$

$$\mu_3' = \frac{q}{p^3}\left(1+4q+q^2\right) \quad (3.5.17)$$

$$\mu_4' = \frac{\partial^4}{\partial t^4}M_X(t)\bigg| t=0$$

$$\frac{\partial^4}{\partial t^4}M_X(t) = \frac{\partial}{\partial t}\left\{\frac{\partial^3}{\partial t^3}M_X(t)\right\}$$

$$\frac{\partial^4}{\partial t^4}M_X(t) = \frac{\partial}{\partial t}\left[pq\left[(1-qe^t)^{-2}e^t + 6qe^{2t}(1-qe^t)^{-3} + 6q^2e^{3t}(1-qe^t)^{-4}\right]\right]$$

$$= pq\left[\frac{\partial}{\partial t}(1-qe^t)^{-2}e^t + \frac{\partial}{\partial t}6qe^{2t}(1-qe^t)^{-3} + \frac{\partial}{\partial t}6q^2e^{3t}(1-qe^t)^{-4}\right]$$

$$= pq\left[\begin{array}{l}(1-qe^t)^{-2}e^t + 2(1-qe^t)^{-3}e^t\left(qe^t\right) + 12qe^{2t}(1-qe^t)^{-3} + 18qe^{2t}(1-qe^t)^{-4}\left(qe^t\right)\\ +18q^2e^{3t}(1-qe^t)^{-4} + 24q^2e^{3t}(1-qe^t)^{-5}\left(qe^t\right)\end{array}\right]$$

$$= pq\left[\begin{array}{l}(1-qe^t)^{-2}e^t + 14qe^{2t}(1-qe^t)^{-3} + 36q^2e^{3t}(1-qe^t)^{-4}\\ +24q^3e^{4t}(1-qe^t)^{-5}\end{array}\right]$$

$$\mu_4' = \left[\frac{\partial^4}{\partial t^4}M_X(t)\bigg| t=0\right] = pq\left[\frac{1}{p^2}+\frac{14q}{p^3}+\frac{36q^2}{p^4}+\frac{24q^3}{p^5}\right]$$

$$= \frac{q}{p^4}\left[p^3+14qp^2+36q^2p+24q^3\right]$$

$$= \frac{q}{p^4}\left[1-3q+3q^2-q^3+14qp^2+36q^2p+24q^3\right]$$

$$= \frac{q}{p^4}\left[1-3q+3q^2-q^3+14qp+22q^2p+24q^3\right]$$

$$= \frac{q}{p^4}\left[1-3q+3q^2-q^3+14qp+22q^2p+22q^3+2q^3\right]$$

$$= \frac{q}{p^4}\left[1-3q+3q^2+q^3+14qp+22q^2\right]$$

$$= \frac{q}{p^4}\left[1-3q+3q^2+q^3+14q-14q^2+22q^2\right]$$

$$\mu_4' = \frac{q}{p^4}\left[1+11q+11q^2+q^3\right] \tag{3.5.18}$$

➢ **Note:** The expression PGF can be derived using the same MGF method. The formula of PGF is $P_X(t) = \sum_x t^x P(X=x)$.

e. **Characteristic Function (CF) of Geometric Distribution:** The expression for CF ($\phi_X(t)$) is obtained as below:

$$\phi_X(t) = E(e^{itX}) = \sum_{x=0} e^{itx} P(X=x)$$

$$= \sum_{x=0} e^{itx} pq^x = p\sum_{x=0} e^{itx} q^x = p\sum_{x=0} (e^{it}q)^x$$

$$= p\frac{1}{1-e^{it}q} \quad \text{[using sum of geometric series]}$$

$$\phi_X(t) = p\left(1-qe^{it}\right)^{-1}.$$

f. **Cumulant Generating Function (CGF) of Geometric Distribution:** The detailed derivation of the CDF is given below:

$$K_X(t) = \log(M_X(t))$$

$$K_X(t) = -\log p\left(1-qe^{it}\right)$$

We can get center moments using cumulant generating function:

$$K_X(t) = \log p(1-qe^t)^{-1}$$

$$= \log p - \log\left(1-q\left\{1+\frac{t}{1!}+\frac{t^2}{2!}+\frac{t^3}{3!}+\frac{t^4}{4!}+....\right\}\right) \quad \text{[using exponential series]}$$

$$= \log p - \log\left(1-q+q\left\{\frac{t}{1!}+\frac{t^2}{2!}+\frac{t^3}{3!}+\frac{t^4}{4!}+....\right\}\right)$$

$$= \log p - \log\left(p + q\left\{\frac{t}{1!} + \frac{t^2}{2!} + \frac{t^3}{3!} + \frac{t^4}{4!} + \ldots.\right\}\right)$$

$$= -\log\left(1 + \frac{q}{p}\left\{\frac{t}{1!} + \frac{t^2}{2!} + \frac{t^3}{3!} + \frac{t^4}{4!} + \ldots.\right\}\right)$$

$$= -\left[\begin{array}{l} -\frac{q}{p}\left\{\frac{t}{1!} + \frac{t^2}{2!} + \frac{t^3}{3!} + \frac{t^4}{4!} + \ldots.\right\} - \frac{1}{2}\left(\frac{q}{p}\right)^2\left\{\frac{t}{1!} + \frac{t^2}{2!} + \frac{t^3}{3!} + \frac{t^4}{4!} + \ldots.\right\}^2 \\ -\frac{1}{3}\left(\frac{q}{p}\right)^3\left\{\frac{t}{1!} + \frac{t^2}{2!} + \frac{t^3}{3!} + \frac{t^4}{4!} + \ldots.\right\}^3 - \frac{1}{4}\left(\frac{q}{p}\right)^4\left\{\frac{t}{1!} + \frac{t^2}{2!} + \frac{t^3}{3!} + \frac{t^4}{4!} + \ldots.\right\}^4 .. \end{array}\right]$$

κ_r = coefficient of $\frac{t^r}{r!}$ in $K_X(t)$

κ_1 = coefficient of $\frac{t}{1!}$ in $K_X(t) = \frac{q}{p}$

κ_2 = coefficient of $\frac{t^2}{2!}$ in $K_X(t) = \frac{q}{p} + \left(\frac{q}{p}\right)^2 = \frac{q}{p^2}$

κ_3 = coefficient of $\frac{t^3}{3!}$ in $K_X(t)$

$$\kappa_3 = \frac{q}{p} + 3\left(\frac{q}{p}\right)^2 + 2\left(\frac{q}{p}\right)^3 = \frac{q}{p^3}\left[p^2 + 3qp + 2q^2\right]$$

$$= \frac{q}{p}\left[1 + q^2 - 2q + 3q - 3q^2 + 2q^2\right] = \frac{q(1+q)}{p^3}$$

Similarly, we get κ_4 = coefficient of $\frac{t^4}{4!}$ in $K_X(t)$.

3.6 NEGATIVE BINOMIAL DISTRIBUTION

Negative Binomial distribution contains a countable number of values (0, 1, 2, …) with success probability (p) and failure probability (q) with two conditions i) $0 \leq p \leq 1$ and ii) $p + q = 1$. For example, the number of loan offers given by the lender to the customers until the 5th loan is sanctioned. This distribution gives the probability of that number of failures before rth success. In Negative Binomial, distribution, the PMF of a random variable X is:

$$P(X=x)=\begin{cases}\dfrac{(x+r-1)!}{(r-1)!(x)!}\mathrm{p}^r\mathrm{q}^x & \mathrm{x}=0,1,2,\ldots\\ 0 & \text{otherwise}\end{cases}$$

- **Another Form of Negative Binomial Distribution:**

$$P(X=x)=\begin{cases}(-1)^x\dfrac{-r!}{(-r-x)!(x)!}\mathrm{p}^r\mathrm{q}^x & \mathrm{x}=0,1,2,\ldots\\ 0 & \text{otherwise}\end{cases}$$

- **Remark:** if $p=\frac{1}{Q}\Rightarrow Q=\frac{1}{p}$ and $q=\frac{P}{Q}\Rightarrow P=\frac{q}{p}$

Since $p+q=\frac{1}{Q}+\frac{P}{Q}\Rightarrow P+1=Q=Q-P=1$

$$P(X=x)=\begin{cases}(-1)^x\,{}^{-r}C_x\left(\dfrac{1}{Q}\right)^r\left(\dfrac{P}{Q}\right)^x & \mathrm{x}=0,1,2,\ldots\\ 0 & \text{otherwise}\end{cases}$$

or

$$P(X=x)=\begin{cases}(-1)^x\,{}^{-r}C_xQ^{-r-x}P^x & \mathrm{x}=0,1,2,\ldots\\ 0 & \text{otherwise}\end{cases} \qquad (3.6.1)$$

- **Sample Generation from Negative Binomial Distribution** (Figure 3.10)

R Code:

$x\ = rnbinom(n, size, p)$
$where$
$n = sample\ size$
$k = number\ of\ successful\ trials$
$p = probability\ of\ success$
$Example$
$x =\ rnbinom(10, 20,0.5)$

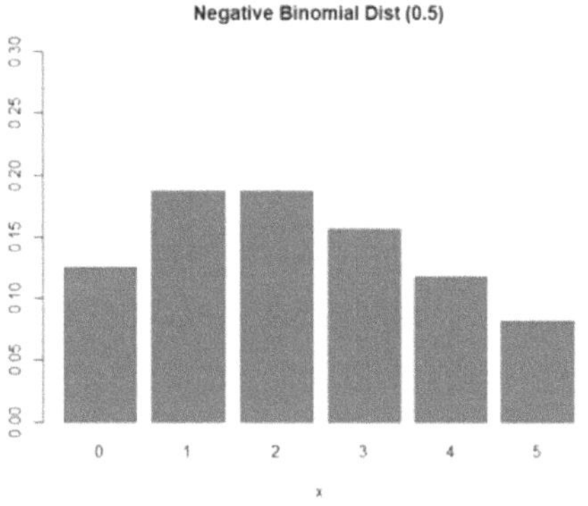

FIGURE 3.10 Negative binomial distribution.

- **Output:**

```
Console   Terminal   Jobs

> x= rnbinom(10, 20,0.5)
> x
 [1] 19  7 31 18 10 21 25 23 27 17
>
```

3.6.1 PROPERTIES OF NEGATIVE BINOMIAL DISTRIBUTION

a. **Moment Generating Function (MGF) of Negative Binomial Distribution:** It is another method for obtaining moments from the probability model. The expression for MGF ($M_X(t)$) is obtained as below:

$$M_X(t) = E(e^{tX}) = \sum_{x=0} e^{tx} P(X = x)$$

$$= \sum_{x=0} e^{tx} (-1)^x \, {}^{-r}C_x Q^{-r-x} P^x = \sum_{x=0} {}^{-r}C_x Q^{-r-x} \left(-e^t P\right)^x$$

$$M_X(t) = (Q - Pe^t)^{-r} \quad \text{[using Binomial theorem]} \quad (3.6.2)$$

We can get moments using MGF:

$$\mu_r' = \frac{\partial^r}{\partial t^r} M_X(t) \bigg| t = 0$$

$$\mu_1' = \frac{\partial}{\partial t} M_X(t) \bigg| t = 0$$

$$\frac{\partial}{\partial t} M_X(t) = \frac{\partial}{\partial t} (Q - Pe^t)^{-r} = (-r)(Q - Pe^t)^{-r-1} \left(-Pe^t\right)$$

$$= rPe^t (Q - Pe^t)^{-r-1} = rP(Q - P)^{-r-1} \quad \text{[using Eqn. (3.6.2)]}$$

$$\frac{\partial}{\partial t} M_X(t) = rP(Q - P)^{-r-1} \quad (3.6.3)$$

$$\mu_1' = \left[\frac{\partial}{\partial t} M_X(t) \bigg| t = 0 \right] = rP = r\frac{q}{p} \quad (3.6.4)$$

$$\mu_2' = \frac{\partial^2}{\partial t^2} M_X(t) \bigg| t = 0$$

$$\frac{\partial^2}{\partial t^2} M_X(t) = \frac{\partial}{\partial t} \left\{ \frac{\partial}{\partial t} M_X(t) \right\}$$

$$= \frac{\partial}{\partial t}\left[(rP)(Q - Pe^t)^{-r-1} e^t\right] = rP\frac{\partial}{\partial t}\left[(Q - Pe^t)^{-r-1} e^t\right] \text{ [using Eqn. (3.6.4)]}$$

$$= rP\left[(Q - Pe^t)^{-r-1} e^t + (-r-1)(Q - Pe^t)^{-r-2}(-Pe^t)e^t\right]$$

$$\frac{\partial^2}{\partial t^2} M_X(t) = rP\left[(Q - Pe^t)^{-r-1} e^t + (r+1)Pe^{2t}(Q - Pe^t)^{-r-2}\right] \qquad (3.6.5)$$

$$\mu_2' = \left[\frac{\partial^2}{\partial t^2} M_X(t)\bigg| t = 0\right] = rP\left[1 + (r+1)P\right]$$

$$\mu_2' = rP\left[1 + (r+1)P\right] = r\frac{q}{p}\left[1 + (r+1)\frac{q}{p}\right] = r\frac{q}{p^2}\left(1 + rq\right) \qquad (3.6.6)$$

$$\mu_3' = \frac{\partial^3}{\partial t^3} M_X(t)\bigg| t = 0$$

$$\frac{\partial^3}{\partial t^3} M_X(t) = \frac{\partial^3}{\partial t^3} M_X(t) = \frac{\partial}{\partial t}\left[\frac{\partial^2}{\partial t^2} M_X(t)\right]$$

$$\frac{\partial^3}{\partial t^3} M_X(t) = \frac{\partial}{\partial t}\left[rP\left[(Q - Pe^t)^{-r-1} e^t + (r+1)Pe^{2t}(Q - Pe^t)^{-r-2}\right]\right]$$

[using Eqn. (3.6.5)]

$$= rP\frac{\partial}{\partial t}\left[(Q - Pe^t)^{-r-1} e^t + (r+1)Pe^{2t}(Q - Pe^t)^{-r-2}\right]$$

$$= rP\left[\frac{\partial}{\partial t}(Q - Pe^t)^{-r-1} e^t + (r+1)P\frac{\partial}{\partial t}e^{2t}(Q - Pe^t)^{-r-2}\right]$$

$$= rP\begin{bmatrix}(Q - Pe^t)^{-r-1} e^t + (r+1)Pe^{2t}(Q - Pe^t)^{-r-2} + 2(r+1)Pe^{2t}(Q - Pe^t)^{-r-2} \\ (r+1)(r+2)Pe^{2t}(Q - Pe^t)^{-r-3}(Pe^t)\end{bmatrix}$$

$$\frac{\partial^3}{\partial t^3} M_X(t) = rP\begin{bmatrix}(Q - Pe^t)^{-r-1} e^t + 3(r+1)Pe^{2t}(Q - Pe^t)^{-r-2} \\ +(r+1)(r+2)P^2 e^{3t}(Q - Pe^t)^{-r-3}\end{bmatrix} \qquad (3.6.7)$$

$$\mu_3' = \left[\frac{\partial^3}{\partial t^3} M_X(t)\bigg| t = 0\right] = rP\left[1 + 3(r+1)P + (r+1)(r+2)P^2\right]$$

$$= r\frac{q}{p}\left[1 + 3(r+1)\frac{q}{p} + (r+1)(r+2)\left(\frac{q}{p}\right)^2\right]$$

$$= r\frac{q}{p}\left[p^2 + 3(r+1)qp + (r+1)(r+2)q^2\right]$$

$$= r\frac{q}{p^3}\left[p^2 + 3qrp + 3qp + r^2q^2 + 3rq^2 + 2q^2\right]$$

$$= r\frac{q}{p^3}\left[1 + q^2 + qp + 3qrp + 3rq^2 + r^2q^2\right]$$

$$\mu_3' = rP\left[1 + 3(r+1)P + (r+1)(r+2)P^2\right] = r\frac{q}{p^3}\left[1 + q + 3qr + r^2q^2\right] \quad (3.6.8)$$

Similarly, we can find $\mu_4' = \frac{\partial^4}{\partial t^4} M_X(t)\Big|\, t = 0$

➢ **Note:** The expression PGF can be derived using the same MGF method. The formula of PGF is $P_X(t) = \sum_x t^x P(X = x)$.

b. **Characteristic Function (CF) of Negative Binomial Distribution:** The expression for CF ($\phi_X(t)$) is obtained as below:

$$\phi_X(t) = E(e^{itX}) = \sum_{x=0} e^{itx} P(X = x)$$

$$= \sum_{x=0} e^{itx} (-1)^x \frac{-r!}{(-r-x)!(x)!} Q^{-r-x} P^x = \sum_{x=0} \frac{-r!}{(-r-x)!(x)!} Q^{-r-x} (-e^{it}P)^x$$

$$K_X(t) = \log(M_X(t)) \quad \text{[using Binomial theorem]}$$

c. **Cumulant Generating Function (CGF) of Negative Binomial Distribution:** We can derive the expression of the CGF to see the highlights of the probability model. The detailed derivation of the CDF is given below:

$$K_X(t) = \log(M_X(t))$$

$$K_X(t) = \log(Q - Pe^t)^{-r}$$

We can get center moments using cumulant generating function:

$$K_X(t) = -r\log(Q - Pe^t)$$

$$= -r\log\left(Q - P\left\{1 + \frac{t}{1!} + \frac{t^2}{2!} + \frac{t^3}{3!} + \frac{t^4}{4!} + \dots\right\}\right) \quad \text{[using exponential series]}$$

$$= -r\log\left(Q - P + P\left\{\frac{t}{1!} + \frac{t^2}{2!} + \frac{t^3}{3!} + \frac{t^4}{4!} + \dots\right\}\right)$$

$$= -r\log\left(1 + P\left\{\frac{t}{1!} + \frac{t^2}{2!} + \frac{t^3}{3!} + \frac{t^4}{4!} + \dots\right\}\right)$$

$$= -r\left[\begin{array}{l} -P\left\{\frac{t}{1!}+\frac{t^2}{2!}+\frac{t^3}{3!}+\frac{t^4}{4!}+....\right\}-\frac{1}{2}P^2\left\{\frac{t}{1!}+\frac{t^2}{2!}+\frac{t^3}{3!}+\frac{t^4}{4!}+....\right\}^2 \\ -\frac{1}{3}P^3\left\{\frac{t}{1!}+\frac{t^2}{2!}+\frac{t^3}{3!}+\frac{t^4}{4!}+....\right\}^3-\frac{1}{4}P^3\left\{\frac{t}{1!}+\frac{t^2}{2!}+\frac{t^3}{3!}+\frac{t^4}{4!}+....\right\}^4.. \end{array}\right]$$

κ_r = coefficient of $\frac{t^r}{r!}$ in $K_X(t)$

κ_1 = coefficient of $\frac{t}{1!}$ in $K_X(t) = rP = r\frac{q}{p}$

κ_2 = coefficient of $\frac{t^2}{2!}$ in $K_X(t) = rP + rP^2 = r\frac{q}{p} + r\left(\frac{q}{p}\right)^2 = \frac{rqp + rq^2}{p^2}$

$$= \frac{rq(p+q)}{p^2} = \frac{rq}{p^2}$$

κ_3 = coefficient of $\frac{t^3}{3!}$ in $K_X(t)$

$$\kappa_3 = rP + 3rP^2 + 2rP^3 = r\frac{q}{p} + 3r\left(\frac{q}{p}\right)^2 + 2r\left(\frac{q}{p}\right)^3$$

$$= r\frac{q}{p}\left[\frac{p^2 + 3qp + 2q^2}{p^2}\right] = r\frac{q}{p}\left[\frac{1 + q^2 + qp}{p^2}\right] = r\frac{q(1+q)}{p^3}$$

Similarly, we get κ_4 = coefficient of $\frac{t^4}{4!}$ in $K_X(t) = \kappa_4 = rPQ\,(1 + 6PQ) = r\frac{q}{p^4}(p^2 + 6q)$

Further, we can find $\mu_4 = \kappa_4 + 3\kappa_2^2 = r\frac{q}{p^4}\left[p^2 + 3q(r+2)\right]$.

- **Note:** The expression PGF can be derived using the same MGF method. The formula of PGF is $P_X(t) = \sum_x t^x P(X = x)$.

d. Skewness and Kurtosis of Negative Binomial Distribution: The expressions of skewness (β_1, γ_1) and kurtosis (β_2, γ_2) four coefficients based upon the first four central moments are given below:

$$\beta_1 = \frac{\mu_3^2}{\mu_2^3} = \frac{\left[r\frac{q(1+q)}{p^3}\right]^2}{\left[r\frac{q}{p^2}\right]^3} = \frac{(1+q)^2}{rq}$$

$$\gamma_1 = \frac{\mu_3}{\mu_2^{3/2}} = \frac{(1+q)}{\sqrt{rq}}$$

$$\beta_2 = \frac{\mu_4}{\mu_2^2} = \frac{r\frac{q}{p^4}\left[p^2 + 3q(r+2)\right]}{\left[r\frac{q}{p^2}\right]^2} = \frac{\left[p^2 + 3q(r+2)\right]}{rq}$$

$$\beta_2 = 3 + \frac{\left[p^2 + 6q\right]}{rq}$$

$$\gamma_2 = \beta_2 - 3 = \frac{\left[p^2 + 6q\right]}{rq}$$

➢ **Note:** The other properties of the Negative Binomial distribution (Figure 3.11) can be derived utilizing the methods used above for the earlier derivations.

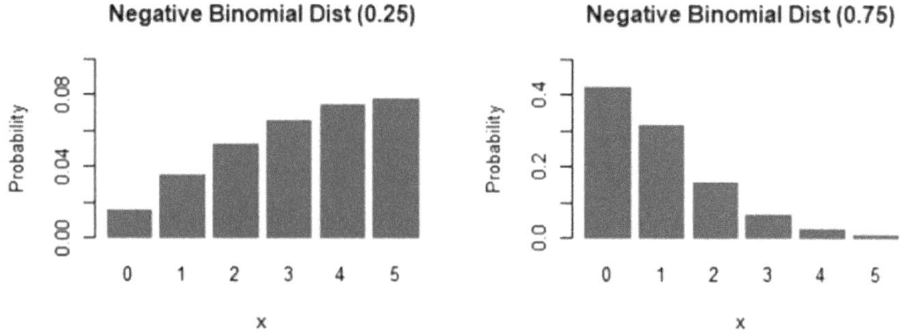

FIGURE 3.11 Shape of the negative binomial distribution.

3.7 HYPER-GEOMETRIC DISTRIBUTION

Hyper-geometric distribution contains a non-negative finite number of values. Let us consider a container with N 'T-shirts,' out of which 'M' are white 'T-shirts' and rest (N-M) are Black 'T-shirts.' Further, we take the sample size n using simple random sampling without replacement. In the sample, the number of white shirts is 'k' and the rest (n-k) are black. The hypergeometric distribution gives the probability of getting 'k' white 'T-shirts' out of the n sample. In Hyper-Geometric distribution, the PMF is:

$$P(X=k)=h(k;N,M,n)=\begin{cases}\dfrac{{}^{M}C_{k}\,{}^{N-M}C_{n-k}}{{}^{N}C_{n}} & k=0,1,2....\min(n,M)\\ 0 & \text{otherwise}\end{cases}$$

Symbolically: X ~ *Hypergeomtric*(*N, M, n, k*).

Here N, M, n, k are parameters of the distribution and $k < n$.

➢ **Sample Generation from Hypergeometric Distribution** (Figure 3.12)

R Code:

```
x = rhyper (nn, m, n, k)
where
nn = sample size
m = number of white balls
n = number of black balls
Example
x = rhyper(10, 10, 7,8)
```

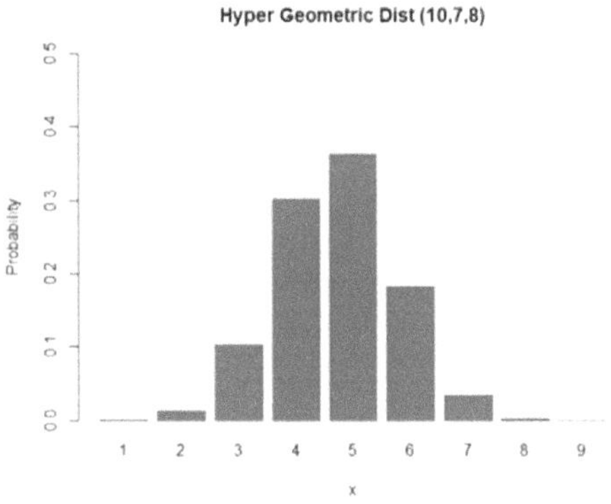

FIGURE 3.12 Hypergeometric distribution.

➢ **Output:**

```
Console  Terminal  Jobs
> x= rhyper(10, 10, 7,8)
> x
[1] 3 6 4 5 5 5 6 6 6 3
>
```

3.7.1 PROPERTIES OF HYPER-GEOMETRIC DISTRIBUTION

a. **Moments of Hyper-Geometric Distribution:** The first four raw moments $\mu_1', \mu_2', \mu_3', \mu_4'$ have been derived as below:

$$\mu_1' = E(X) = \sum_{x=1}^{n} xP(X=x)$$

$$= \sum_{x=0}^{n} x\frac{{}^{M}C_{x}\,{}^{N-M}C_{n-x}}{{}^{N}C_{n}}$$

$$= \sum_{x=1}^{n} \left[\frac{\frac{M!}{x-1!(M-x)!} \frac{(N-M)!}{(n-x)!(N-M-n+x)!}}{\frac{N!}{n!(N-n)!}} \right]$$

$$= M \frac{1}{\frac{N!}{n!(N-n)!}} \sum_{x=1}^{n} \left[\frac{M-1!}{x-1!(M-x)!} \frac{(N-M)!}{(n-x)!(N-M-n+x)!} \right]$$

$$= M \frac{1}{\frac{N!}{n!(N-n)!}} \sum_{x=1}^{n} \left[\frac{M-1!}{x-1!(M-x)!} \frac{(N-M)!}{(n-x)!(N-M-n+x)!} \right]$$

$$\begin{bmatrix} let \quad x-1=k \\ when \ \mathrm{x} = 1 \Rightarrow \mathrm{k} = 0 \\ \mathrm{x} = \mathrm{n} \Rightarrow \mathrm{k} = \mathrm{n}\text{-}1 \end{bmatrix}$$

$$= M \frac{1}{\frac{N!}{n!(N-n)!}} \sum_{k=0}^{n-1} \left[\frac{M-1!}{k!(M-k-1)!} \frac{(N-M)!}{(n-k-1)!(N-M-(n-1-k))!} \right]$$

$$= M \frac{1}{\frac{N!}{n!(N-n)!}} \sum_{k=0}^{n-1} \left[\frac{M-1!}{k!(M-1-k)!} \frac{(N-M)!}{(n-1-k)!(N-M-(n-1-k))!} \right]$$

Since we have $\frac{a+b!}{r!(a+b-r)!} = \sum_{k=0}^{r} \frac{a!}{k!(a-k)!} \frac{b!}{(r-k)!(b-r+k)!}$

$$= M \frac{1}{\frac{N!}{n!(N-n)!}} \frac{\{(M-1)+(N-M)\}!}{\{(M-1)+(N-M)-(n-1)\}!(n-1)!}$$

$$= M \frac{1}{\frac{N!}{n!(N-n)!}} \frac{(N-1)!}{(N-n)(n-1)!}$$

We know that $\frac{N!}{n!(N-n)!} = \frac{N}{n} \frac{(N-1)!}{\{(N-1)-(n-1)\}(n-1)!}$

$$= M \frac{1}{\frac{N!}{n!(N-n)!}} \frac{(N-1)!}{(N-n)(n-1)!}$$

$$= M \frac{1}{\frac{N}{n}\frac{(N-1)!}{(N-n)(n-1)!}} \frac{(N-1)!}{(N-n)(n-1)!}$$

$$\mu_1' = \frac{n}{N} M \qquad (3.7.1)$$

Thus, $\frac{n}{N} M$ is the mean of the distribution.

$$\mu_2' = E(X^2) = \sum_{x=0}^{n} x^2 P(X = x)$$

$$= \sum_{x=0}^{n} x^2 \frac{{}^M C_x \, {}^{N-M} C_{n-x}}{{}^N C_n}$$

$$= \sum_{x=0}^{n} [x(x-1) + x] \frac{{}^M C_x \, {}^{N-M} C_{n-x}}{{}^N C_n}$$

$$= \sum_{x=0}^{n} x(x-1) \frac{{}^M C_x \, {}^{N-M} C_{n-x}}{{}^N C_n} + \sum_{x=0}^{n} x \frac{{}^M C_x \, {}^{N-M} C_{n-x}}{{}^N C_n}$$

$$= \sum_{x=0}^{n} x(x-1) \left[\frac{{}^M C_x \, {}^{N-M} C_{n-x}}{{}^N C_n} \right] + \mu_1' \qquad \text{[using Eqn. (3.7.1)]}$$

$$= \sum_{x=0}^{n} x(x-1) \left[\frac{{}^M C_x \, {}^{N-M} C_{n-x}}{{}^N C_n} \right] + \frac{n}{N} M$$

$$= \sum_{x=2}^{n} \left[\frac{\frac{M!}{x-2!(M-x)!} \frac{(N-M)!}{(n-x)!(N-M-n+x)!}}{\frac{N!}{n!(N-n)!}} \right] + \frac{n}{N} M$$

$$= M(M-1) \sum_{x=2}^{n} \left[\frac{\frac{M-2!}{x-2!(M-x)!} \frac{(N-M)!}{(n-x)!(N-M-n+x)!}}{\frac{N!}{n!(N-n)!}} \right] + \frac{n}{N} M$$

$$\left[\begin{array}{l} let \quad x-2 = k \\ when \ \mathrm{x} = 2 \Rightarrow \mathrm{k} = 0 \\ \mathrm{x} = \mathrm{n} \Rightarrow \mathrm{k} = \mathrm{n\text{-}2} \end{array} \right]$$

$$= M(M-1)\sum_{k=0}^{n-2}\left[\frac{\frac{M-2!}{x-2!(M-2-k)!}\frac{(N-M)!}{(n-2-k)!(N-M-n+2+k)!}}{\frac{N!}{n!(N-n)!}}\right]+\frac{n}{N}M$$

Since we have $\frac{a+b!}{r!(a+b-r)!}=\sum_{k=0}^{r}\frac{a!}{k!(a-k)!}\frac{b!}{(r-k)!(b-r+k)!}$

$$= M(M-1)\frac{\frac{N-2!}{n-2!(N-n)!}}{\frac{N!}{n!(N-n)!}}+\frac{n}{N}M$$

$$\mu_2' = M(M-1)\frac{n(n-1)}{N(N-1)}+\frac{n}{N}M \tag{3.7.2}$$

$$\mu_3' = E(X^3)=\sum_{x=0}^{n}x^3P(X=x)$$

$$=\sum_{x=0}^{n}x^3\frac{{}^{M}C_x\,{}^{N-M}C_{n-x}}{{}^{N}C_n}$$

$$=\sum_{x=0}^{n}\left[x(x-1)(x-2)+3x(x-1)+x\right]\frac{{}^{M}C_x\,{}^{N-M}C_{n-x}}{{}^{N}C_n}$$

$$=\left[\begin{array}{l}\sum_{x=0}^{n}x(x-1)(x-2)\frac{{}^{M}C_x\,{}^{N-M}C_{n-x}}{{}^{N}C_n}+\sum_{x=0}^{n}3x(x-1)\frac{{}^{M}C_x\,{}^{N-M}C_{n-x}}{{}^{N}C_n}\\+\sum_{x=0}^{n}x\frac{{}^{M}C_x\,{}^{N-M}C_{n-x}}{{}^{N}C_n}\end{array}\right]$$

$$=\sum_{x=0}^{n}x(x-1)(x-2)\frac{{}^{M}C_x\,{}^{N-M}C_{n-x}}{{}^{N}C_n}+3\mu_2'+\mu_1' \quad \text{[using (3.7.1) and (3.7.2)]}$$

$$=\sum_{x=0}^{n}\frac{\frac{M!}{x-3!(M-x)!}\frac{(N-M)!}{(n-x)!(N-M-n+x)!}}{\frac{N!}{n!(N-n)!}}+3\frac{n(n-1)M(M-1)}{N(N-1)}+\frac{nM}{N}$$

$$= M(M-1)(M-2)\sum_{x=3}^{n}\frac{\frac{M-3!}{x-3!(M-x)!}\frac{(N-M)!}{(n-x)!(N-M-n+x)!}}{\frac{N!}{n!(N-n)!}}$$

$$+3\frac{n(n-1)M(M-1)}{N(N-1)}+\frac{nM}{N}$$

$$\begin{bmatrix} let \quad x-3=k \\ when\ \mathrm{x}=3 \Rightarrow \mathrm{k}=0 \\ \mathrm{x}=\mathrm{n} \Rightarrow \mathrm{k}=\mathrm{n\text{-}3} \end{bmatrix}$$

$$= M(M-1)(M-2)\sum_{k=0}^{n-3}\frac{\frac{M-3!}{k!(M-3-k)!}\frac{(N-M)!}{(n-3-k)!(N-M-(n-3-k))!}}{\frac{N!}{n!(N-n)!}}$$

$$+3\frac{n(n-1)M(M-1)}{N(N-1)}+\frac{nM}{N}$$

Since we have $\frac{a+b!}{r!(a+b-r)!}=\sum_{k=0}^{r}\frac{a!}{k!(a-k)!}\frac{b!}{(r-k)!(b-r+k)!}$

$$= M(M-1)(M-2)\frac{\frac{N-3!}{n-3!(N-n)!}}{\frac{N!}{n!(N-n)!}}+3\frac{n(n-1)M(M-1)}{N(N-1)}+\frac{nM}{N}$$

$$= M(M-1)(M-2)\frac{n(n-1)(n-2)}{N(N-1)(N-2)}+3\frac{n(n-1)M(M-1)}{N(N-1)}+\frac{nM}{N}$$

$$\mu_3' = n(n-1)(n-2)\frac{M(M-1)(M-2)}{N(N-1)(N-2)}+3n(n-1)\frac{M(M-1)}{N(N-1)}+n\frac{M}{N} \quad (3.7.3)$$

$$\mu_4' = E(X^4) = \sum_{x=0}^{n} x^4 P(X=x)$$

$$= \sum_{x=0}^{n} x^4 \frac{{}^{M}C_x\, {}^{N-M}C_{n-x}}{{}^{N}C_n}$$

$$= \sum_{x=0}^{n} [x(x-1)(x-2)(x-3) + 6x(x-1)(x-2) + 7x(x-1) + x] \frac{{}^{M}C_{x}\,{}^{N-M}C_{n-x}}{{}^{N}C_{n}}$$

$$= \left[\begin{array}{l} \sum_{x=0}^{n} x(x-1)(x-2)(x-3) \frac{{}^{M}C_{x}\,{}^{N-M}C_{n-x}}{{}^{N}C_{n}} + \sum_{x=0}^{n} 6x(x-1)(x-2) \frac{{}^{M}C_{x}\,{}^{N-M}C_{n-x}}{{}^{N}C_{n}} \\ + \sum_{x=0}^{n} 7x(x-1) \frac{{}^{M}C_{x}\,{}^{N-M}C_{n-x}}{{}^{N}C_{n}} + \sum_{x=0}^{n} x \frac{{}^{M}C_{x}\,{}^{N-M}C_{n-x}}{{}^{N}C_{n}} \end{array} \right]$$

$$= \sum_{x=0}^{n} x(x-1)(x-2)(x-3) \frac{{}^{M}C_{x}\,{}^{N-M}C_{n-x}}{{}^{N}C_{n}} + 6\mu_3' + 7\mu_2' + \mu_1'$$

$$= \sum_{x=4}^{n} \frac{\frac{M!}{x-4!(M-x)!} \frac{(N-M)!}{(n-x)!(N-M-n+x)!}}{\frac{N!}{n!(N-n)!}} + 6\mu_3' + 7\mu_2' + \mu_1'$$

$$= M(M-1)(M-2)(M-3) \sum_{x=4}^{n} \frac{\frac{M-4!}{x-4!(M-x)!} \frac{(N-M)!}{(n-x)!(N-M-n+x)!}}{\frac{N!}{n!(N-n)!}}$$

$$+6\mu_3' + 7\mu_2' + \mu_1'$$

$$\left[\begin{array}{l} let \quad x-4=k \\ when\ \mathrm{x}=4 \Rightarrow \mathrm{k}=0 \\ \mathrm{x}=\mathrm{n} \Rightarrow \mathrm{k}=\mathrm{n}\text{-}4 \end{array} \right]$$

$$= M(M-1)(M-2)(M-3) \sum_{k=0}^{n-4} \frac{\frac{M-4!}{k!(M-4-k)!} \frac{(N-M)!}{(n-4-k)!(N-M-n+4+k)!}}{\frac{N!}{n!(N-n)!}}$$

$$+6\mu_3' + 7\mu_2' + \mu_1'$$

Since we have $\frac{a+b!}{r!(a+b-r)!} = \sum_{k=0}^{r} \frac{a!}{k!(a-k)!} \frac{b!}{(r-k)!(b-r+k)!}$

$$= M(M-1)(M-2)(M-3) \frac{\frac{N-4!}{(n-4)!(N-n)!}}{\frac{N!}{n!(N-n)!}} + 6\mu_3' + 7\mu_2' + \mu_1'$$

$$= M(M-1)(M-2)(M-3)\frac{n(n-1)(n-2)(n-3)}{N(N-1)(N-2)(N-3)} + 6\mu_3' + 7\mu_2' + \mu_1'$$

$$= \begin{bmatrix} n(n-1)(n-2)(n-3)\dfrac{M(M-1)(M-2)(M-3)}{N(N-1)(N-2)(N-3)} \\ +6n(n-1)(n-2)\dfrac{M(M-1)(M-2)}{N(N-1)(N-2)} + 7n(n-1)\dfrac{M(M-1)}{N(N-1)} + n\dfrac{M}{N} \end{bmatrix}$$

$$\mu_4' = n\frac{M}{N}\left[1 + (n-1)\frac{(M-1)}{(N-1)}\left\{7 + (n-2)\frac{(M-2)}{(N-2)}\left(6 + (n-3)\frac{(M-3)}{(N-3)}\right)\right\}\right] \quad (3.7.4)$$

b. **Central Moments Hyper-Geometric Distribution:** The first four central moments μ_1, μ_2, μ_3, μ_4 have been derived as below:

$$\mu_1 = 0$$

$$\mu_2 = \mu'_2 - \mu'^2_1$$

$$= n(n-1)\frac{M(M-1)}{N(N-1)} + n\frac{M}{N} + \left[n\frac{M}{N}\right]^2 \text{ [using Eqns. (3.7.1) and (3.7.2)]}$$

$$= n\frac{M}{N}\left[(n-1)\frac{(M-1)}{(N-1)} + 1 - n\frac{M}{N}\right]$$

$$= n\frac{M}{N^2(N-1)}\left[N(n-1)(M-1) + N(N-1) - nM(N-1)\right]$$

$$= n\frac{M}{N^2(N-1)}\left[NnM - Nn - NM + N + N^2 - N - nNM + nM\right]$$

$$= n\frac{M}{N^2(N-1)}\left[-Nn - NM + N^2 + nM\right]$$

$$\mu_2 = n\frac{M}{N^2(N-1)}\text{(N-n)(N-M)} \quad (3.7.5)$$

So, the *variance* of the *Hyper-geometric* distribution is $n\frac{M}{N^2(N-1)}$ (N-n) (N-M)

Similarly, we get third and fourth central moments.

- **Note:** The other properties of the Hypergeometric distribution can be derived utilizing the methods used above for the earlier derivations.

KEYWORDS

- **characteristic function**
- **hyper-geometric distribution**
- **moment generating function**
- **probability generating function**
- **probability mass function**
- **random variable**

REFERENCES

Casella, G., & Berger, R. L., (2002). *Statistical Inference*. Duxbury: Belmont, CA.

Gupta, S. C., & Kapoor, V. K., (1997). *Fundamentals of Mathematical Statistics*. Sultan Chand and Sons: New Delhi.

Heumann, C., Schomaker, M., & Shalabh, (2016). *Introduction to Statistics and Data Analysis: With Exercises, Solutions and Applications in R*. Springer.

Hogg, R. V., McKean, J. W., & Craig, A. T., (2013). *Introduction to Mathematical Statistics* (7th edn.). Pearson: Boston.

Miller, I., Miller, M., Freund, J. E., & Miller, I., (2004). *John E. Freund's Mathematical Statistics with Applications*. Prentice Hall: Upper Saddle River, NJ.

Rohatgi, V., Saleh, A. K. Md. E., (2015). *Introduction to Probability and Statistics* (3rd edn.). John Wiley.

CHAPTER 4

Continuous Distributions

SHWETA DIXIT[1], MUKTI KHETAN[2], MOHD. ARSHAD[3], PRASHANT VERMA[4], and ASHOK KUMAR PATHAK[5]

[1]*Clinical Development Services Agency, Translational Health Science and Technology Institute, NCR Biotech Science Cluster, Haryana, India*

[2]*Department of Mathematics, Amity School of Applied Sciences, Amity University Mumbai, Maharashtra, India*

[3]*Department of Mathematics, Indian Institute of Technology Indore, Simrol, Indore, Madhya Pradesh, India*

[4]*Department of Statistics, Faculty of Science, University of Allahabad, Prayagraj, India*

[5]*Department of Mathematics and Statistics, School of Basic and Applied Sciences, Central University of Punjab, Bathinda, Punjab, India*

4.1 UNIFORM DISTRIBUTION

A uniform distribution is defined on an interval (a, b). If a random variable X follows a uniform distribution, the expression of the probability density function (PDF) is (Figure 4.1):

$$f(x;a,b) = \begin{cases} \dfrac{1}{b-a} & \text{if } \ a < x < b \\ 0 & \text{otherwise} \end{cases}$$

Distribution Theory: Principles and Applications. Fozia Homa, Mukti Khetan, Mohd. Arshad, & Pradeep Mishra (Naz) (Eds.)

Symbolically: X~U (a, b).

Here 'a' and 'b' are the parameters of the distribution.
The cumulative distribution function (CDF) is:

$$F(x) = P(X \leq x) = \int_a^x f(t)dt = \int_a^x \frac{1}{b-a}dt = \frac{x-a}{b-a}, \quad a \leq x<b.$$

$$F(x) = \begin{cases} 0 & x < a \\ \dfrac{x-a}{b-a} & a \leq x < b \\ 1 & x \geq b \end{cases}$$

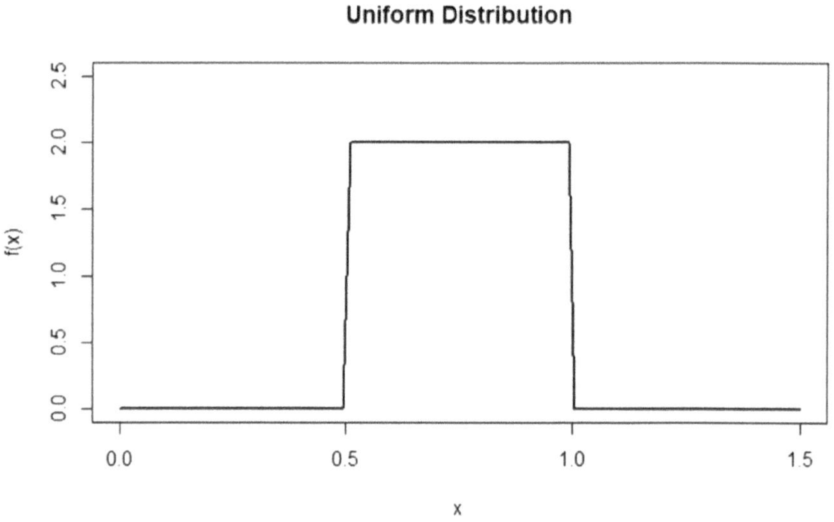

FIGURE 4.1 Uniform distribution.

➢ Sample Generation of Uniform Distribution

R Code: Generate a random sample from a continuous uniform distribution.

x = runif(n, b, a)

where; n is the sample size; a, b are the range of the distribution

Example:

x = runif(10,0,1)

Output:

```
Console  Terminal  Jobs
> x=runif(10,0,1)
> x
 [1] 0.43236132 0.66021913 0.53509049 0.26218147 0.09583048 0.89810587 0.75752595 0.66696252 0.87695364
[10] 0.58230323
> |
```

4.1.1 PROPERTIES OF UNIFORM DISTRIBUTION

a. **Raw Moments of Uniform Distribution:** The first four raw moments $\mu'_1, \mu'_2, \mu'_3, \mu'_4$ have been derived as below:

$$\mu_1' = E(X) = \int_x xf(x)dx = \int_a^b xf(x)dx = \int_a^b x\frac{1}{b-a}dx = \frac{1}{b-a}\left[\frac{b^2-a^2}{2}\right].$$

$$\mu_1' = \frac{(a+b)}{2}. \qquad (4.1.1)$$

So, *mean* of the uniform distribution is $\frac{(a+b)}{2}$.

$$\mu_2' = E(X^2) = \int_a^b x^2 f(x)dx = \int_a^b x^2\frac{1}{b-a}dx = \frac{1}{b-a}\left[\frac{b^3-a^3}{3}\right]$$

$$= \frac{\left(a^2+b^2+ab\right)}{3}. \qquad (4.1.2)$$

$$\mu_3' = E(X^3) = \int_a^b x^3\frac{1}{b-a}dx = \frac{1}{b-a}\left[\frac{b^4-a^4}{4}\right] = \frac{1}{b-a}\frac{\left(a^2+b^2\right)(a+b)(b-a)}{4}$$

$$= \frac{\left(a^2+b^2\right)(a+b)}{4}. \qquad (4.1.3)$$

$$\mu_4' = E(X^4) = \int_a^b x^4\frac{1}{b-a}dx = \left[\frac{b^5-a^5}{5}\right]$$

$$= \frac{1}{b-a}\frac{(b-a)\left(a^4+a^3b+a^2b^2+ab^3+b^4\right)}{5}$$

$$= \frac{\left(a^4+a^3b+a^2b^2+ab^3+b^4\right)}{5}. \qquad (4.1.4)$$

b. **Central Moments of Uniform Distribution:** The first four central moments μ_1, μ_2, μ_3, μ_4 have been derived as below:

$$\mu_1 = 0 \text{ [first central moment is always zero]} \qquad (4.1.5)$$

$$\mu_1 = \mu_2' - \mu_1'^2$$

$$= \frac{\left(a^2+b^2+ab\right)}{3} - \left[\frac{(a+b)}{2}\right]^2 = \frac{1}{12}\left[4a^2+4b^2+4ab-3a^2-3b^2-6ab\right]$$

[using Eqns. (4.1.1) and (4.1.2)]

$$\mu_2 = \frac{1}{12}(b-a)^2. \qquad (4.1.6)$$

So, *variance* of the uniform distribution is $\frac{1}{12}(b-a)^2$

$$\mu_3 = \mu_3' - 3\mu_2'\,\mu_1', + 2\,\mu_1'^3$$

$$= \frac{\left(a^2+b^2\right)(a+b)}{4} - 3\frac{\left(a^2+b^2+ab\right)}{3}\frac{(a+b)}{2} + 2\left[\frac{(a+b)}{2}\right]^3$$

[using Eqns. (4.1.1)–(4.1.3)]

$$= \frac{(a+b)}{4}\left\{\left(a^2+b^2\right) - 2\left(a^2+b^2+ab\right) + (a+b)^2\right\}$$

$$= 0 \qquad (4.1.7)$$

$$\mu_4 = \mu_4' - 4\mu_3'\mu_1' + 6\mu_2'\mu_1'^2 - 3\mu_1'^4$$

$$= \frac{\left(a^4+a^3b+a^2b^2+ab^3+b^4\right)}{5} - 4\frac{\left(a^2+b^2\right)(a+b)}{4}\frac{(a+b)}{2}$$
$$+6\frac{\left(a^2+b^2+ab\right)}{3}\left[\frac{(a+b)}{2}\right]^2 - 3\left[\frac{(a+b)}{2}\right]^4$$

[Using Eqns. (4.1.1)–(4.1.4)]

$$= \frac{\left(a^4+a^3b+a^2b^2+ab^3+b^4\right)}{5} - 4\frac{\left(a^2+b^2\right)(a+b)}{4}\frac{(a+b)}{2}$$
$$+\frac{\left(a^2+b^2+ab\right)(a+b)^2}{2} - 3\frac{(a+b)^4}{16}$$

$$=\frac{\left(a^4+a^3b+a^2b^2+ab^3+b^4\right)}{5}-4\frac{\left(a^2+b^2\right)(a+b)}{4}\frac{(a+b)}{2}+\frac{(a+b)^2}{16}\left[8\left(a^2+b^2+ab\right)-3(a+b)^2\right]$$

$$=\frac{\left(a^4+a^3b+a^2b^2+ab^3+b^4\right)}{5}-4\frac{\left(a^2+b^2\right)(a+b)}{4}\frac{(a+b)}{2}+\frac{(a+b)^2}{16}\left(5a^2+5b^2+2ab\right)$$

$$=\frac{\left(a^4+a^3b+a^2b^2+ab^3+b^4\right)}{5}+\frac{(a+b)^2}{16}\left[-8\left(a^2+b^2\right)+\left(5a^2+5b^2+2ab\right)\right]$$

$$=\frac{\left(a^4+a^3b+a^2b^2+ab^3+b^4\right)}{5}+\frac{(a+b)^2}{16}\left[-3a^2-3b^2+2ab\right]$$

$$=\frac{\left(a^4+a^3b+a^2b^2+ab^3+b^4\right)}{5}+\frac{(-3a^4-2a^2b^2-4ab^3-4a^3b-3b^4)}{16}$$

$$=\frac{1}{80}\begin{bmatrix}16a^4+16a^3b+16a^2b^2+16ab^3+16b^4-15a^4\\ -10a^2b^2-20ab^3-20a^3b-15b^4\end{bmatrix}$$

$$=\frac{1}{80}[b-a]^4. \qquad (4.1.8)$$

c. **Skewness and Kurtosis of Uniform Distribution:** To find the shape of the distribution, we have derived the expressions of skewness (β_1, γ_1) and kurtosis (β_2, γ_2), which are given below:

$$\beta_1=\frac{\mu_3^2}{\mu_2^3}=0. \quad \text{[using Eqns. (4.1.6) and (4.1.7)]}$$

$$\gamma_1=\sqrt{\beta_1}=\frac{\mu_3}{\mu_2^{3/2}}=0.$$

$$\beta_2=\frac{\mu_4}{\mu_2^2}=\frac{1}{80}\frac{(b-a)^4}{\left[\frac{1}{12}(b-a)^2\right]^2}. \quad \text{[using Eqns. (4.1.6) and (4.1.8)]}$$

$$\beta_2=1.8$$

$$\gamma_2=\beta_2-3=-1.2.$$

This implies that distribution is always symmetric and platykurtic.

d. **Moment Generating Function (MGF) of Uniform Distribution:** The expression for MGF ($M_X(t)$) is obtained as below:

$$M_X(t) = E(e^{tX}) = \int_a^b e^{tx} f(x)dx = \int_a^b e^{tx} \frac{1}{b-a} dx = \frac{1}{b-a}\left[\frac{e^{tx}}{t}\right]_a^b = \frac{1}{b-a}\left[\frac{e^{tb}-e^{ta}}{t}\right]$$

$$\Rightarrow M_X(t) = \begin{cases} \frac{1}{t(b-a)}\left[e^{tb}-e^{ta}\right] & ; t \neq 0, \\ 1 & ; t = 0 \end{cases} \quad a \neq b.$$

e. **Characteristic Function (CF) of Uniform Distribution:** The expression for CF is obtained as below:

$$\phi_X(t) = E(e^{itX}) = \int_a^b e^{itx} f(x)dx = \int_a^b e^{itx} \frac{1}{b-a} dx$$

$$\Rightarrow \phi_X(t) = \begin{cases} \frac{1}{it(b-a)}\left[e^{itb}-e^{ita}\right]; & t \neq 0, \\ 1 & ; t = 0 \end{cases} \quad a \neq b.$$

Note: The expression PGF can be derived using the same MGF method. The formula of PGF is $P_X(t) = \int_x t^x f(x)dx.$

4.2 GAMMA DISTRIBUTION

The gamma probability model is defined in the range of $(0, \infty)$. If a random variable X follows gamma distribution, the expression of the PDF is:

$$f_X(x) = \begin{cases} \frac{1}{(\alpha-1)!}\lambda^\alpha e^{-\lambda x} x^{\alpha-1} & 0 < x < \infty \\ 0 & \text{otherwise} \end{cases}$$

Symbolically: $X \sim G(\alpha, \lambda)$

Here α, λ are the parameters of the distribution and $\alpha, \lambda > 0$.
The CDF of the gamma distribution is (Figure 4.2):

$$F_X(x) = \int_{-\infty}^x f(t)dt = \int_0^x f(t)dt = \int_0^x \frac{1}{(\alpha-1)!}\lambda^\alpha e^{-\lambda t} t^{\alpha-1} dt \quad ; 0 < x < \infty$$

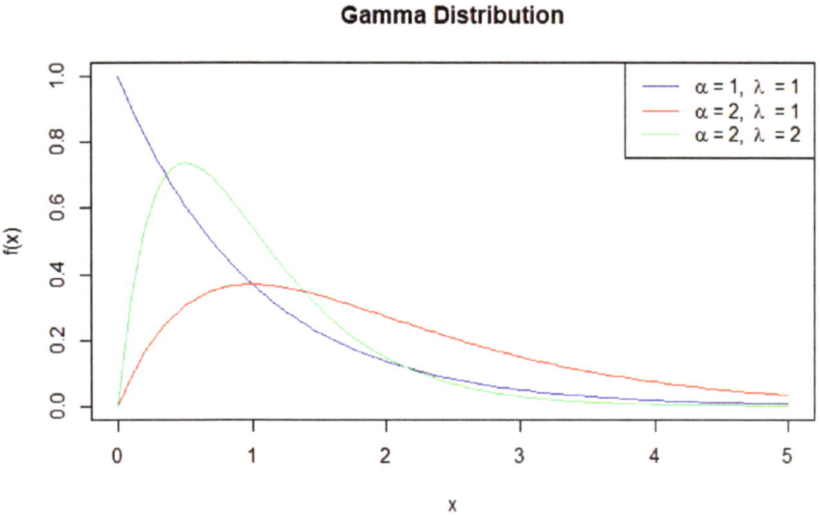

FIGURE 4.2 Gamma distribution.

➢ **Sample Generation of Gamma Distribution**

R Code: Generate a random sample from a gamma distribution.

x = rgamma (n, α, λ)

where; n is the sample size; α, λ are the parameters of the distribution

Example:

x = rgamma (10, 1, 1)

Output:

```
Console   Terminal   Jobs

> x=rgamma(10, 1, 1)
> x
 [1] 0.93260694 1.12112052 1.10467751 0.96000269 1.21886061 0.07426174 1.39423637 0.16397753 0.34904599
[10] 1.54696869
>
```

4.2.1 *PROPERTIES OF GAMMA DISTRIBUTION*

a. **Raw Moments of Gamma Distribution:** The first four raw moments $\mu_1', \mu_2', \mu_3', \mu_4'$ have been derived as below:

$$\mu_1' = E(X) = \int_0^\infty x f(x)dx = \int_0^\infty x \frac{1}{(\alpha-1)!}\lambda^\alpha e^{-\lambda x} x^{\alpha-1}\, dx = \frac{1}{(\alpha-1)!}\lambda^\alpha \int_0^\infty x^\alpha e^{-\lambda x} dx$$

$$Let\ \ \lambda x = y \Rightarrow dx = \frac{1}{\lambda}dy \quad \& \, x = \frac{y}{\lambda}$$

$$= \frac{1}{(\alpha-1)!}\lambda^\alpha \int_0^\infty \left(\frac{y}{\lambda}\right)^\alpha e^{-y}\frac{1}{\lambda}dy = \frac{1}{(\alpha-1)!}\frac{1}{\lambda}\int_0^\infty y^\alpha e^{-y} dy$$

Using gamma function, we get:

$$\mu_1' = \frac{\alpha}{\lambda}. \qquad (4.2.1)$$

So, *mean* of the gamma distribution is $\frac{\alpha}{\lambda}$.

$$\mu_2' = E(X^2) = \int_0^\infty x^2 f(x)dx = \int_0^\infty x^2 \frac{1}{(\alpha-1)!}\lambda^\alpha e^{-\lambda x} x^{\alpha-1}\, dx = \frac{1}{(\alpha-1)!}\lambda^\alpha \int_0^\infty x^{\alpha+1} e^{-\lambda x} dx$$

$$Let\ \ \lambda x = y \Rightarrow dx = \frac{1}{\lambda}dy \quad \& \, x = \frac{y}{\lambda}$$

$$= \frac{1}{(\alpha-1)!}\lambda^\alpha \int_0^\infty \left(\frac{y}{\lambda}\right)^{\alpha+1} e^{-y}\frac{1}{\lambda}dy = \frac{1}{(\alpha-1)!}\frac{1}{\lambda^2}\int_0^\infty y^{\alpha+1} e^{-y} dy,$$

Using gamma function, we get:

$$\mu_2' = \frac{\alpha(\alpha+1)}{\lambda^2}. \qquad (4.2.2)$$

$$\mu_3' = E(X^3) = \int_0^\infty x^3 f(x)dx = \int_0^\infty x^3 \frac{1}{(\alpha-1)!}\lambda^\alpha e^{-\lambda x} x^{\alpha-1}\, dx = \frac{1}{(\alpha-1)!}\lambda^\alpha \int_0^\infty x^{\alpha+2} e^{-\lambda x} dx$$

$$Let\ \ \lambda x = y \Rightarrow dx = \frac{1}{\lambda}dy \quad \& \, x = \frac{y}{\lambda}$$

$$= \frac{1}{(\alpha-1)!}\lambda^\alpha \int_0^\infty \left(\frac{y}{\lambda}\right)^{\alpha+2} e^{-y}\frac{1}{\lambda}dy = \frac{1}{(\alpha-1)!}\frac{1}{\lambda^3}\int_0^\infty y^{\alpha+2} e^{-y} dy,$$

Using gamma function, we get:

$$\mu_3' = \frac{\alpha(\alpha+1)(\alpha+2)}{\lambda^3}. \qquad (4.2.3)$$

$$\mu_4' = E(X^4) = \int_0^\infty x^4 f(x)dx = \int_0^\infty x^4 \frac{1}{(\alpha-1)!}\lambda^\alpha e^{-\lambda x} x^{\alpha-1}\, dx = \frac{1}{(\alpha-1)!}\lambda^\alpha \int_0^\infty x^{\alpha+3} e^{-\lambda x} dx$$

$$Let\ \ \lambda x = y \Rightarrow dx = \frac{1}{\lambda}dy \quad \& \, x = \frac{y}{\lambda}$$

$$\frac{1}{(\alpha-1)!}\lambda^{\alpha}\int_{0}^{\infty}\left(\frac{y}{\lambda}\right)^{\alpha+3} e^{-y}\frac{1}{\lambda}dy = \frac{1}{(\alpha-1)!}\frac{1}{\lambda^4}\int_{0}^{\infty} y^{\alpha+3}e^{-y}dy = \frac{1}{(\alpha-1)!}\frac{1}{\lambda^4}(\alpha+3)!,$$

Using gamma function, we get:

$$\mu_4' = \frac{\alpha(\alpha+1)(\alpha+2)(\alpha+3)}{\lambda^4}. \tag{4.2.4}$$

- **Alternative Method:** We can also find μ_r' and get all four moments

$$\mu_r' = E(X^r) = \int_{0}^{\infty} x^r f(x)dx = \int_{0}^{\infty} x^r \frac{1}{(\alpha-1)!}\lambda^{\alpha}e^{-\lambda x}x^{\alpha-1}\, dx = \frac{1}{(\alpha-1)!}\lambda^{\alpha}\int_{0}^{\infty} x^{\alpha+r-1}e^{-\lambda x}dx$$

$$Let\ \ \lambda x = y \Rightarrow dx = \frac{1}{\lambda}dy \quad \& \, x = \frac{y}{\lambda}$$

$$= \frac{1}{(\alpha-1)!}\lambda^{\alpha}\int_{0}^{\infty}\left(\frac{y}{\lambda}\right)^{\alpha+r-1} e^{-y}\frac{1}{\lambda}dy = \frac{1}{(\alpha-1)!}\frac{1}{\lambda^r}\int_{0}^{\infty} y^{\alpha+r-1}e^{-y}dy = \frac{1}{(\alpha-1)!}\frac{1}{\lambda^r}(\alpha+r-1)!,$$

Using gamma function, we get:

$$\mu_r' = \frac{(\alpha+r-1)!}{(\alpha-1)!}\frac{1}{\lambda^r}. \tag{4.2.5}$$

Substitute $r = 1, 2, 3, 4$ in Eqn. (4.2.5) and get all four moments μ_1', μ_2', μ_3', μ_4' which are mentioned in Eqns. (4.2.1) to (4.2.4) respectively.

b. Central Moments of Gamma Distribution: The first four central moments μ_1, μ_2, μ_3, μ_4 have been derived as below:

$$\mu_1 = 0 \tag{4.2.6}$$

$$\mu_2 = \mu_2' - \mu_1'^2 = \frac{\alpha(\alpha+1)}{\lambda^2} - \left(\frac{\alpha}{\lambda}\right)^2 \quad \text{[using Eqns. (4.2.1) and (4.2.2)]}$$

$$\mu_2 = \frac{\alpha}{\lambda^2}. \tag{4.2.7}$$

So, *variance* of the gamma distribution is $\frac{\alpha}{\lambda^2}$

$$\mu_3 = \mu_3' - 3\mu_2'\mu_1' + 2\mu_1'^3 = \frac{\alpha(\alpha+1)(\alpha+2)}{\lambda^3} - 3\frac{\alpha(\alpha+1)}{\lambda^2}\frac{\alpha}{\lambda} + 2\left(\frac{\alpha}{\lambda}\right)^3$$

[using Eqns. (4.2.1)–(4.2.3)]

$$=\frac{\alpha}{\lambda^3}\left[(\alpha+1)(\alpha+2)-3\alpha(\alpha+1)+2\alpha^2\right]=\frac{\alpha}{\lambda^3}\left[\alpha^2+3\alpha+2-3\alpha^2-3\alpha+2\alpha^2\right]$$

$$\mu_3=\frac{2\alpha}{\lambda^3}. \tag{4.2.8}$$

$$\mu_4=\mu_4'-4\mu_3'\mu_1'+6\mu_2'\mu_1'^2-3\mu_1'^4$$

$$=\frac{\alpha(\alpha+1)(\alpha+2)(\alpha+3)}{\lambda^4}-4\frac{\alpha(\alpha+1)(\alpha+2)}{\lambda^3}\frac{\alpha}{\lambda}+6\frac{\alpha(\alpha+1)}{\lambda^2}\left(\frac{\alpha}{\lambda}\right)^2-3\left(\frac{\alpha}{\lambda}\right)^4$$

[using Eqns. (4.2.1)–(4.2.4)]

$$=\frac{\alpha\left[(\alpha+1)(\alpha+2)(\alpha+3)-4\alpha(\alpha+1)(\alpha+2)+6\alpha^2(\alpha+1)-3\alpha^3\right]}{\lambda^4}$$

$$=\frac{\alpha\left[(\alpha^3+6\alpha^2+11\alpha+6)-4(\alpha^3+3\alpha^2+2\alpha)+6(\alpha^3+\alpha^2)-3\alpha^3\right]}{\lambda^4}$$

$$\mu_4=\frac{3\alpha(\alpha+2)}{\lambda^4}. \tag{4.2.9}$$

c. **Skewness and Kurtosis of Gamma Distribution:** To find the shape of the distribution, we have derived the expressions of skewness (β_1, γ_1) and kurtosis (β_2, γ_2), which are given below:

$$\beta_1=\frac{\mu_3^2}{\mu_2^3}=\frac{\left(\frac{2\alpha}{\lambda^3}\right)^2}{\left(\frac{\alpha}{\lambda^2}\right)^3}$$

$$\beta_1=\frac{4}{\alpha}$$

$$\gamma_1=\sqrt{\beta_1}=\frac{\mu_3}{\mu_2^{3/2}}=\sqrt{\frac{4}{\alpha}}=2\sqrt{\frac{1}{\alpha}}$$

$$\beta_2=\frac{\mu_4}{\mu_2^2}=\frac{\frac{3\alpha(\alpha+2)}{\lambda^4}}{\left[\frac{\alpha}{\lambda^2}\right]^2}$$

$$\beta_2=\frac{3(\alpha+2)}{\alpha}$$

$$\gamma_2 = \beta_2 - 3 = \frac{3(\alpha+2)}{\alpha} - 3 = \frac{6}{\alpha}.$$

This implies that gamma distribution is always positively skewed and leptokurtic. Note that as $\alpha \to \infty$, this distribution approaches to symmetric and mesokurtic curves.

d. **Moment Generating Function (MGF) of Gamma Distribution:** The expression for MGF ($M_X(t)$) is obtained as below:

$$M_X(t) = E(e^{tX}) = \int_0^\infty e^{tx} f(x)dx = \int_0^\infty e^{tx} \frac{1}{(\alpha-1)!} \lambda^\alpha e^{-\lambda x} x^{\alpha-1}\, dx$$

$$= \int_0^\infty e^{tx} \frac{1}{(\alpha-1)!} \lambda^\alpha e^{-\lambda x} x^{\alpha-1}\, dx = \frac{1}{(\alpha-1)!} \lambda^\alpha \int_0^\infty e^{-x(\lambda-t)} x^{\alpha-1}\, dx, \quad t < \lambda$$

$$Let\ (\lambda - t)x = y \Rightarrow dx = \frac{1}{\lambda - t} dy \quad \& \ x = \frac{y}{\lambda - t}$$

$$= \frac{1}{(\alpha-1)!} \lambda^\alpha \int_0^\infty e^{-y} \left(\frac{y}{(\lambda-t)}\right)^{\alpha-1} \frac{1}{(\lambda-t)}\, dy = \frac{1}{(\alpha-1)!} \frac{\lambda^\alpha}{(\lambda-t)^\alpha} \int_0^\infty y^{\alpha-1} e^{-y} dy,$$

Using gamma function, we get:

$$M_X(t) = \left(1 - \frac{t}{\lambda}\right)^{-\alpha}, \quad t < \lambda$$

We can get moments using MGF.

Using $(1 - x)^{-n}$ expansion, we get:

$$M_X(t) = \left(1 - \frac{t}{\lambda}\right)^{-\alpha} = \left\{ \begin{array}{l} 1 + \frac{(\alpha)}{1!}\left(\frac{t}{\lambda}\right) + \frac{\alpha(\alpha+1)}{2!}\left(\frac{t}{\lambda}\right)^2 + \frac{(\alpha)(\alpha+1)(\alpha+2)}{3!}\left(\frac{t}{\lambda}\right)^3 \\ + \frac{(\alpha)(\alpha+1)(\alpha+2)(\alpha+3)}{4!}\left(\frac{t}{\lambda}\right)^4 + \ldots. \end{array} \right\}.$$

$$\mu_r' = \text{coefficient of } \frac{t^r}{r!} \text{ in } M_X(t).$$

Substitute $r = 1, 2, 3, 4$ and solve them; we can get all fours moments $\mu_1', \mu_2', \mu_3', \mu_4'$

$$\mu_1' = \text{coefficient of } \frac{t}{1!} \text{ in } M_X(t) = \frac{\alpha}{\lambda}.$$

$$\mu_2' = \text{coefficient of } \frac{t^2}{2!} \text{ in } M_X(t) = \frac{\alpha(\alpha+1)}{\lambda^2}.$$

$$\mu_3' = \text{coefficient of } \frac{t^3}{3!} \text{ in } M_X(t) = \frac{\alpha(\alpha+1)(\alpha+2)}{\lambda^3}.$$

$$\mu_4' = \text{coefficient of } \frac{t^4}{4!} \text{ in } M_X(t) = \frac{\alpha(\alpha+1)(\alpha+2)(\alpha+3)}{\lambda^4}.$$

- **Note:** The expression PGF can be derived using the same MGF method. The formula of PGF is $P_X(t) = \int_x t^x f(x)dx.$

e. **Characteristic Function (CF) of Gamma Distribution:** The expression for CF ($\phi_X(t)$) is obtained as below:

$$\phi_X(t) = E(e^{itX}) = \int_0^\infty e^{itx} f(x)dx = \int_0^\infty e^{itx} \frac{1}{(\alpha-1)!} \lambda^\alpha e^{-\lambda x} x^{\alpha-1}\, dx$$

$$= \int_0^\infty e^{itx} \frac{1}{(\alpha-1)!} \lambda^\alpha e^{-\lambda x} x^{\alpha-1}\, dx = \frac{1}{(\alpha-1)!} \lambda^\alpha \int_0^\infty e^{-x(\lambda-it)} x^{\alpha-1}\, dx$$

Note that the integral $\int_0^\infty e^{-x(\lambda-it)} x^{\alpha-1}\, dx$ is a complex integral and the solution of this integral is beyond the scope of this book. After solving this complex integral, we get:

$$\phi_X(t) = \left(1 - \frac{it}{\lambda}\right)^{-\alpha}.$$

f. **Cumulants Generating Function (CGF) of Gamma Distribution:** The expression for CGF ($K_X(t)$) is obtained as below:

$$K_X(t) = \log(M_X(t))$$

$$K_X(t) = \log\left(1 - \frac{t}{\lambda}\right)^{-\alpha}.$$

We can get central moments using cumulants generating function:

$$\kappa_1 = \text{mean}; \quad \kappa_2 = \text{variance}; \quad \kappa_3 = \mu_3; \quad \kappa_4 = -3\kappa_2^2 + \mu_4.$$

$$K_X(t) = \log\left(1 - \frac{t}{\lambda}\right)^{-\alpha} = -\alpha \log\left(1 - \frac{t}{\lambda}\right)$$

$$= -\alpha\left\{-\frac{1}{1}\frac{t}{\lambda} - \frac{1}{2}\left(\frac{t}{\lambda}\right)^2 - \frac{1}{3}\left(\frac{t}{\lambda}\right)^3 - \frac{1}{4}\left(\frac{t}{\lambda}\right)^4 + \ldots\right\} \text{ [using log(1-x) expansion]}$$

$$K_X(t) = \alpha\left\{\frac{1}{1!}\frac{t}{\lambda} + \frac{1}{2!}\left(\frac{t}{\lambda}\right)^2 + \frac{2}{3!}\left(\frac{t}{\lambda}\right)^3 + \frac{6}{4!}\left(\frac{t}{\lambda}\right)^4 + \ldots\right\}.$$

$$\kappa_r = \text{coefficient of } \frac{t^r}{r!} \text{ in } K_X(t).$$

$$\kappa_1 = \text{mean} = \text{coefficient of } \frac{t}{1!} \text{ in } K_X(t) = \frac{\alpha}{\lambda}.$$

$$\kappa_2 = \text{variance} = \text{coefficient of } \frac{t^2}{2!} \text{ in } K_X(t) = \frac{\alpha}{\lambda^2}.$$

$$\kappa_3 = \mu_3 = \text{coefficient of } \frac{t^3}{3!} \text{ in } K_X(t) = \frac{2\alpha}{\lambda^3}.$$

$$\kappa_4 = \text{coefficient of } \frac{t^4}{4!} \text{ in } K_X(t) = \frac{6\alpha}{\lambda^4}.$$

$$\mu_4 = \kappa_4 + 3\kappa_2^2 = \frac{6\alpha}{\lambda^4} + 3\left(\frac{\alpha}{\lambda^2}\right)^2 = \frac{3\alpha(2+\alpha)}{\lambda^4}$$

Note: We can find the expressions of $\mu_1', \mu_2', \mu_3', \mu_4'$ (using Eqn. (4.2.5)).

4.3 BETA DISTRIBUTION OF FIRST KIND

The beta probability model of the first kind is defined in the interval (0,1). If a random variable X follows the beta distribution of the first kind, the expression of the PDF is:

$$f(x) = \begin{cases} \frac{1}{B(\alpha,\beta)} x^{\alpha-1}(1-x)^{\beta-1} & ; \ 0 < x < 1 \\ 0 & \text{otherwise} \end{cases}$$

Symbolically: $X \sim \beta_1(\alpha, \beta)$.
Here α, β are the parameters of the distribution and $\alpha, \beta > 0$.
The CDF of the beta distribution of the first kind is (Figure 4.3):

$$F_X(x) = \begin{cases} 0 & ; \ x < 0 \\ \int_0^x \frac{1}{B(\alpha,\beta)} x^{\alpha-1}(1-x)^{\beta-1} dx & ; \ 0 \le x < 1 \\ 1 & ; \ x \ge 1 \end{cases}$$

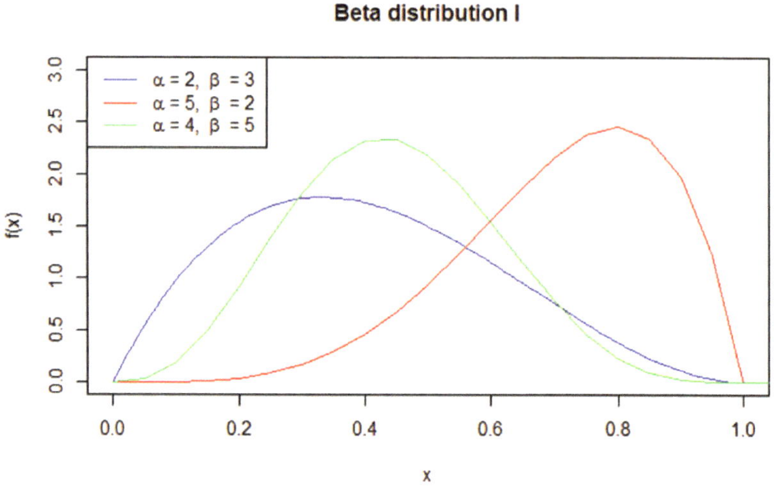

FIGURE 4.3 Beta distribution I.

➢ Sample Generation of Beta Distribution of First Kind

R Code: Generate a random sample from a beta distribution.

x = rbeta(n, α, β)

where; n is the sample size; α, β are the parameters of the distribution

Example:

x = rbeta(10, 2, 3)

Output:

```
Console  Terminal  Jobs
> x=rbeta(10 , 2, 3)
> x
 [1] 0.37404710 0.30155791 0.43806810 0.28312252 0.34341733 0.48073183 0.02733329 0.18808679 0.65350156
[10] 0.47243499
> |
```

4.4 PROPERTIES OF BETA DISTRIBUTION OF FIRST KIND

a. **Raw Moments of Beta Distribution of First Kind:** The first four raw moments $\mu_1', \mu_2', \mu_3', \mu_4'$ have been derived as below:

$$\mu_1' = E(X) = \int_0^1 x \frac{1}{B(\alpha,\beta)} x^{\alpha-1}(1-x)^{\beta-1} dx = \frac{1}{B(\alpha,\beta)} \int_0^1 x^{\alpha}(1-x)^{\beta-1} dx$$

$$= \frac{1}{B(\alpha,\beta)} B(\alpha+1,\beta) \qquad \text{[using Beta function]}$$

$$= \frac{\alpha}{\alpha+\beta}. \tag{4.3.1}$$

So, the **mean** of the beta distribution of the first kind is $\frac{\alpha}{\alpha+\beta}$.

$$\mu_2' = E(X^2) = \int_0^1 x^2 \frac{1}{B(\alpha,\beta)} \mathrm{x}^{\alpha\text{-}1}(1\text{-x})^{\beta\text{-}1}\mathrm{dx} = \frac{1}{B(\alpha,\beta)} \int_0^1 \mathrm{x}^{\alpha+1}(1\text{-x})^{\beta\text{-}1}\mathrm{dx}$$

Using a beta function, we get:

$$= \frac{1}{B(\alpha,\beta)} B(\alpha+2,\beta)$$

$$= \frac{\alpha(\alpha+1)}{(\alpha+\beta+1)(\alpha+\beta)}. \tag{4.3.2}$$

$$\mu_3' = E(X^3) = \int_0^1 x^3 \frac{1}{B(\alpha,\beta)} x^{\alpha\text{-}1}(1-x)^{\beta\text{-}1}dx = \frac{1}{B(\alpha,\beta)} \int_0^1 x^{\alpha+2}(1-x)^{\beta\text{-}1}dx$$

$$= \frac{B(\alpha+3,\beta)}{B(\alpha,\beta)} \qquad \text{[using beta function]}$$

$$= \frac{\alpha(\alpha+1)(\alpha+2)}{(\alpha+\beta)(\alpha+\beta+1)(\alpha+\beta+2)} \tag{4.3.3}$$

$$\mu_4' = E(X^4) = \int_0^1 x^4 \frac{1}{B(\alpha,\beta)} \mathrm{x}^{\alpha\text{-}1}(1\text{-x})^{\beta\text{-}1}\mathrm{dx} = \frac{1}{B(\alpha,\beta)} \int_0^1 \mathrm{x}^{\alpha+3}(1\text{-x})^{\beta\text{-}1}\mathrm{dx}$$

$$= \frac{1}{B(\alpha,\beta)} B(\alpha+4,\beta) \qquad \text{[Using Beta function]}$$

$$= \frac{\alpha(\alpha+1)(\alpha+2)(\alpha+3)}{(\alpha+\beta)(\alpha+\beta+1)(\alpha+\beta+2)(\alpha+\beta+3)} \tag{4.3.4}$$

- **Alternative Method:**

$$\mu_r' = E(X^r) = \int_0^1 x^r \frac{1}{B(\alpha,\beta)} \mathrm{x}^{\alpha\text{-}1}(1\text{-x})^{\beta\text{-}1}\mathrm{dx} = \frac{1}{B(\alpha,\beta)} \int_0^1 \mathrm{x}^{\alpha+r-1}(1\text{-x})^{\beta\text{-}1}\mathrm{dx}$$

$$= \frac{1}{B(\alpha,\beta)} B(\alpha+r,\beta) \qquad \text{[Using Beta function]}$$

$$\mu_r' = \frac{(\alpha+r-1)!(\alpha+\beta-1)!}{(\alpha-1)!(\alpha+\beta+r-1)!}. \tag{4.3.5}$$

Put r = 1 in Eqn. (4.3.5), we get:

$$\mu_1' = \frac{(\alpha)!(\alpha+\beta-1)!}{(\alpha-1)!(\alpha+\beta)!} = \frac{\alpha}{\alpha+\beta}.$$

Put r = 2 in Eqn. (4.3.5), we get:

$$\mu_2' = \frac{(\alpha+1)!(\alpha+\beta-1)!}{(\alpha-1)!(\alpha+\beta+1)!} = \frac{\alpha(\alpha+1)}{(\alpha+\beta)(\alpha+\beta+1)}.$$

Put r = 3 in Eqn. (4.3.5), we get:

$$\mu_3' = \frac{(\alpha+2)!(\alpha+\beta-1)!}{(\alpha-1)!(\alpha+\beta+2)!} = \frac{\alpha(\alpha+1)(\alpha+2)}{(\alpha+\beta)(\alpha+\beta+1)(\alpha+\beta+2)}.$$

Put r = 4 in Eqn. (4.3.5), we get:

$$\mu_4' = \frac{(\alpha+3)!(\alpha+\beta-1)!}{(\alpha-1)!(\alpha+\beta+3)!} = \frac{\alpha(\alpha+1)(\alpha+2)(\alpha+3)}{(\alpha+\beta)(\alpha+\beta+1)(\alpha+\beta+2)(\alpha+\beta+3)}$$

b. **Central Moments of Beta Distribution of First Kind:** The first four central moments $\mu_1, \mu_2, \mu_3, \mu_4$ have been derived as below:

$$\mu_1 = 0 \tag{4.3.6}$$

$$\mu_2 = \mu_2' - \mu_1'^2 = \frac{\alpha(\alpha+1)}{(\alpha+\beta)(\alpha+\beta+1)} - \left(\frac{\alpha}{\alpha+\beta}\right)^2$$

[using Eqns. (6.4.1) and (6.4.2)]

$$= \frac{\alpha}{(\alpha+\beta)^2(\alpha+\beta+1)}\left[(\alpha+1)(\alpha+\beta) - \alpha(\alpha+\beta+1)\right]$$

$$= \frac{\alpha\beta}{(\alpha+\beta)^2(\alpha+\beta+1)} \tag{4.3.7}$$

Thus, a **variance** of the beta distribution of first kind is

$$\frac{\alpha\beta}{(\alpha+\beta)^2(\alpha+\beta+1)}.$$

$$\mu_3 = \mu_3' - 3\mu_2'\mu_1' + 2\mu_1'^3 = \frac{\alpha(\alpha+1)(\alpha+2)}{(\alpha+\beta)(\alpha+\beta+1)(\alpha+\beta+2)}$$

$$-3\frac{\alpha(\alpha+1)}{(\alpha+\beta)(\alpha+\beta+1)}\left(\frac{\alpha}{\alpha+\beta}\right)+2\left(\frac{\alpha}{\alpha+\beta}\right)^3$$

[Using Eqns. (4.3.1)–(4.3.3)]

$$=\frac{\alpha(\alpha+1)(\alpha+2)}{(\alpha+\beta)(\alpha+\beta+1)(\alpha+\beta+2)}+\frac{\alpha^2}{(\alpha+\beta)^2}\left\{-3\frac{(\alpha+1)}{(\alpha+\beta+1)}+2\frac{\alpha}{\alpha+\beta}\right\}$$

$$=\frac{\alpha(\alpha+1)(\alpha+2)}{(\alpha+\beta)(\alpha+\beta+1)(\alpha+\beta+2)}+\frac{\alpha^2}{(\alpha+\beta+1)(\alpha+\beta)^3}$$
$$\left\{-3(\alpha+1)(\alpha+\beta)+2\alpha(\alpha+\beta+1)\right\}$$

$$=\frac{\alpha(\alpha+1)(\alpha+2)}{(\alpha+\beta)(\alpha+\beta+1)(\alpha+\beta+2)}+\frac{\alpha^2}{(\alpha+\beta+1)(\alpha+\beta)^3}$$
$$\left\{-3(\alpha^2+\alpha+\alpha\beta+\beta)+2\alpha(\alpha+\beta+1)\right\}$$

$$=\frac{\alpha(\alpha+1)(\alpha+2)}{(\alpha+\beta)(\alpha+\beta+1)(\alpha+\beta+2)}+\frac{\alpha^2}{(\alpha+\beta+1)(\alpha+\beta)^3}$$
$$\left\{-3(\alpha^2+\alpha+\alpha\beta+\beta)+2(\alpha^2+\alpha\beta+\alpha)\right\}$$

$$=\frac{\alpha(\alpha+1)(\alpha+2)}{(\alpha+\beta)(\alpha+\beta+1)(\alpha+\beta+2)}+\frac{\alpha^2}{(\alpha+\beta+1)(\alpha+\beta)^3}\left\{-\alpha^2-\alpha-\alpha\beta-3\beta\right\}$$

$$=\frac{\alpha}{(\alpha+\beta)(\alpha+\beta+1)}\left[\frac{(\alpha+1)(\alpha+2)}{(\alpha+\beta+2)}+\frac{\alpha\left(-\alpha^2-\alpha-\alpha\beta-3\beta\right)}{(\alpha+\beta)^2}\right]$$

$$=\frac{\alpha}{(\alpha+\beta)^3(\alpha+\beta+1)(\alpha+\beta+2)}$$
$$\left[(\alpha+1)(\alpha+2)(\alpha+\beta)^2+\alpha\left(-\alpha^2-\alpha-\alpha\beta-3\beta\right)(\alpha+\beta+2)\right]$$

$$=\frac{2\alpha\beta(\beta-\alpha)}{(\alpha+\beta)^3(\alpha+\beta+1)(\alpha+\beta+2)}. \tag{4.3.8}$$

Similarly, we get the fourth moment using the formula:

$$\mu_4 = \mu_4' - 4\mu_3'\mu_1' + 6\mu_2'\mu_1'^2 - 3\mu_1'^4$$
$$= \frac{3\alpha\beta\{\alpha\beta(\alpha+\beta-6)+2(\alpha+\beta)^2\}}{(\alpha+\beta)^4(\alpha+\beta+1)(\alpha+\beta+2)(\alpha+\beta+3)} \quad (4.3.9)$$

c. **Skewness and Kurtosis of Beta Distribution of First Kind:** To find the shape of the distribution, we have derived the expressions of skewness (β_1, γ_1) and kurtosis (β_2, γ_2), which are given below:

$$\beta_1 = \frac{\mu_3^2}{\mu_2^3} = \frac{\left[\frac{2\alpha\beta(\beta-\alpha)}{(\alpha+\beta)^3(\alpha+\beta+1)(\alpha+\beta+2)}\right]^2}{\left[\frac{\alpha\beta}{(\alpha+\beta)^2(\alpha+\beta+1)}\right]^3} = \frac{4(\beta-\alpha)^2(\alpha+\beta+1)}{\alpha\beta(\alpha+\beta+2)^2}$$

[using Eqns. (4.3.7) and (4.3.8)]

$$\gamma_1 = \sqrt{\beta_1} = \sqrt{\frac{4(\beta-\alpha)^2(\alpha+\beta+1)}{\alpha\beta(\alpha+\beta+2)^2}} = \frac{2(\beta-\alpha)}{(\alpha+\beta+2)}\sqrt{\frac{(\alpha+\beta+1)}{\alpha\beta}} \quad (4.3.10)$$

$$\beta_2 = \frac{\mu_4}{\mu_2^2} = \frac{\frac{3\alpha\beta\{\alpha\beta(\alpha+\beta-6)+2(\alpha+\beta)^2\}}{(\alpha+\beta)^4(\alpha+\beta+1)(\alpha+\beta+2)(\alpha+\beta+3)}}{\left[\frac{\alpha\beta}{(\alpha+\beta)^2(\alpha+\beta+1)}\right]^2}$$
$$= \frac{3\{\alpha\beta(\alpha+\beta-6)+2(\alpha+\beta)^2\}(\alpha+\beta+1)}{\alpha\beta(\alpha+\beta+2)(\alpha+\beta+3)}$$

[using Eqns. (4.4.7) and (4.4.9)]

$$\gamma_2 = \beta_2 - 3$$

$$= 3\left[\frac{\{\alpha\beta(\alpha+\beta-6)+2(\alpha+\beta)^2\}(\alpha+\beta+1)}{\alpha\beta(\alpha+\beta+2)(\alpha+\beta+3)} - 1\right] \quad (4.3.11)$$

4.5 BETA DISTRIBUTION OF SECOND KIND

Beta distribution of the second kind is defined on the interval (0, ∞). If a random variable X follows the beta distribution of the second kind, the expression of the PDF is (Figure 4.4):

$$f(x)=\begin{cases}\dfrac{1}{B(\alpha,\beta)}\dfrac{x^{\alpha-1}}{(1+x)^{\alpha+\beta}} ; & 0<x<\infty\\ 0 & \text{otherwise}\end{cases}$$

Symbolically: $X \sim \beta_2(\alpha, \beta)$.

Here α, β are the parameters of the distribution and $\alpha, \beta > 0$.

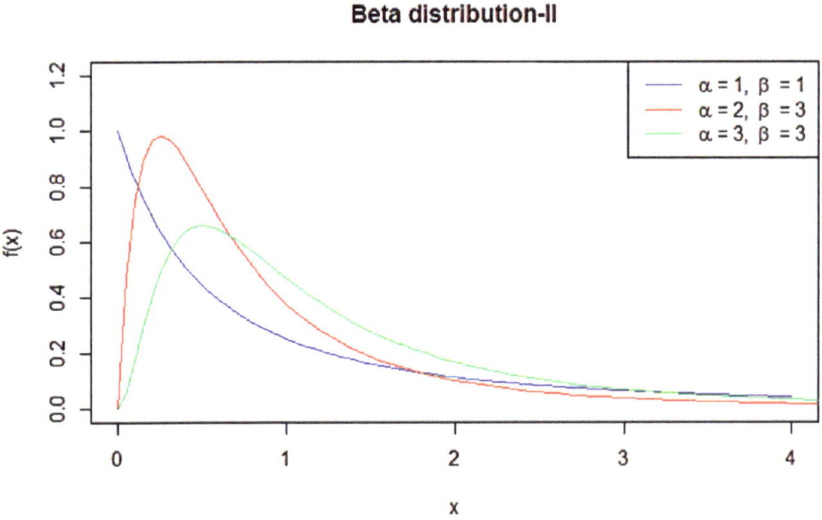

FIGURE 4.4 Beta distribution II.

The CDF of the beta distribution of the second kind is:

$$F_X(x)=\begin{cases}0 & ; x<0\\ \displaystyle\int_0^x \frac{1}{B(\alpha,\beta)}\frac{x^{\alpha-1}}{(1+x)^{\alpha+\beta}}dx & ; 0\le x<1\\ 1 & ; x\ge 1\end{cases}$$

4.5.1 PROPERTIES OF BETA DISTRIBUTION OF SECOND KIND

a. **Moments of Beta Distribution of Second Kind:** The first four central moments μ_1, μ_2, μ_3, μ_4 have been derived as below:

$$\mu_1' = E(X) = \int_0^\infty x \frac{1}{B(\alpha,\beta)} \frac{x^{\alpha-1}}{(1+x)^{\alpha+\beta}} dx = \frac{1}{B(\alpha,\beta)} \int_0^\infty \frac{x^{\alpha}}{(1+x)^{\alpha+\beta}} dx$$

$$= \frac{1}{B(\alpha,\beta)} B(\alpha+1,\beta-1) \qquad \text{[using beta function]}$$

$$= \frac{\alpha}{\beta-1}; \quad \beta > 1. \qquad (4.4.1)$$

So, the **mean** of the beta distribution of the second kind is $\frac{\alpha}{\beta-1}$.

$$\mu_2' = E(X^2) = \int_0^\infty x^2 \frac{1}{B(\alpha,\beta)} \frac{x^{\alpha-1}}{(1+x)^{\alpha+\beta}} dx = \frac{1}{B(\alpha,\beta)} \int_0^\infty \frac{x^{\alpha+1}}{(1+x)^{\alpha+\beta}} dx$$

$$= \frac{1}{B(\alpha,\beta)} B(\alpha+2,\beta-2); \quad \beta > 2 \text{ [using beta function]}$$

$$\mu_2' = \frac{\alpha(\alpha+1)}{(\beta-1)(\beta-2)}; \quad \beta > 2. \qquad (4.4.2)$$

$$\mu_3' = E(X^3) = \int_0^\infty x^3 \frac{1}{B(\alpha,\beta)} \frac{x^{\alpha-1}}{(1+x)^{\alpha+\beta}} dx = \frac{1}{B(\alpha,\beta)} \int_0^\infty \frac{x^{\alpha+2}}{(1+x)^{\alpha+\beta}} dx$$

$$= \frac{1}{B(\alpha,\beta)} B(\alpha+3,\beta-3); \quad \beta > 3 \text{ [using beta function]}$$

$$= \frac{\alpha(\alpha+1)(\alpha+2)}{(\beta-1)(\beta-2)(\beta-3)}; \beta > 3 \qquad (4.4.3)$$

$$\mu_4' = E(X^4) = \int_0^\infty x^4 \frac{1}{B(\alpha,\beta)} \frac{x^{\alpha-1}}{(1+x)^{\alpha+\beta}} dx = \frac{1}{B(\alpha,\beta)} \int_0^\infty \frac{x^{\alpha+3}}{(1+x)^{\alpha+\beta}} dx$$

$$= \frac{1}{B(\alpha,\beta)} B(\alpha+4,\beta-4); \quad \beta > 4 \text{ [Using beta function]}$$

$$= \frac{\alpha(\alpha+1)(\alpha+2)(\alpha+3)}{(\beta-1)(\beta-2)(\beta-3)(\beta-4)}; \beta > 4. \qquad (4.4.4)$$

➢ **Alternative Method:**

$$\mu_r' = E(X^r) = \int_0^\infty x^r \frac{1}{B(\alpha,\beta)} \frac{x^{\alpha-1}}{(1+x)^{\alpha+\beta}} dx = \frac{1}{B(\alpha,\beta)} \int_0^\infty \frac{x^{\alpha+r-1}}{(1+x)^{\alpha+\beta}} dx$$

$$= \frac{1}{B(\alpha,\beta)} B(\alpha+r, \beta-r); \quad \beta > r \text{ [using beta function]}$$

$$\mu_r' = \frac{(\alpha+r-1)!(\beta-r-1)!}{(\alpha-1)!(\beta-1)!}; \quad \beta > r. \tag{4.4.5}$$

Put r = 1 in Eqn. (4.4.5), we get:

$$\mu_1' = \frac{(\alpha)!(\beta-2)!}{(\alpha-1)!(\beta-1)!} = \frac{\alpha}{(\beta-1)}.$$

Put r = 2 in Eqn. (4.4.5), we get:

$$\mu_2' = \mu_r' = \frac{(\alpha+1)!(\beta-3)!}{(\alpha-1)!(\beta-1)!} = \frac{\alpha(\alpha+1)}{(\beta-1)(\beta-2)}.$$

Put r = 3 in Eqn. (4.4.5), we get:

$$\mu_3' = \frac{(\alpha+2)!(\beta-4)!}{(\alpha-1)!(\beta-1)!} = \frac{\alpha(\alpha+1)(\alpha+2)}{(\beta-1)(\beta-2)(\beta-3)}.$$

Put r = 4 in Eqn. (4.4.5), we get:

$$\mu_4' = \frac{(\alpha+3)!(\beta-5)!}{(\alpha-1)!(\beta-1)!} = \frac{\alpha(\alpha+1)(\alpha+2)(\alpha+3)}{(\beta-1)(\beta-2)(\beta-3)(\beta-4)}$$

b. **Central Moments of Beta Distribution of Second Kind:** The first four central moments μ_1, μ_2, μ_3, μ_4 have been derived as below:

$$\mu_1 = 0 \tag{4.4.6}$$

$$\mu_2 = \mu_2' - \mu_1'^2 = \frac{\alpha(\alpha+1)}{(\beta-1)(\beta-2)} - \left(\frac{\alpha}{(\beta-1)}\right)^2 \text{ [using Eqns. (4.4.1) and (4.4.2)]}$$

$$= \frac{\alpha}{(\beta-1)}\left[\frac{(\alpha+1)}{(\beta-2)} - \left(\frac{\alpha}{(\beta-1)}\right)\right]$$

$$= \frac{\alpha(\beta+\alpha-1)}{(\beta-1)^2(\beta-2)} \tag{4.4.7}$$

So, **variance** of the beta distribution of second kind is $\frac{\alpha(\beta+\alpha-1)}{(\beta-1)^2(\beta-2)}; \beta > 2.$

$$\mu_3 = \mu_3' - 3\mu_2'\mu_1' + 2\mu_1'^3 = \frac{\alpha(\alpha+1)(\alpha+2)}{(\beta-1)(\beta-2)(\beta-3)}$$

$$-3\frac{\alpha(\alpha+1)}{(\beta-1)(\beta-2)}\left(\frac{\alpha}{(\beta-1)}\right)+2\left(\frac{\alpha}{(\beta-1)}\right)^3 \qquad \text{[using Eqns. (4.4.1)–(4.4.3)]}$$

$$= \frac{\alpha(\alpha+1)(\alpha+2)}{(\beta-1)(\beta-2)(\beta-3)} + \frac{\alpha^2}{(\beta-1)^2}\left\{-3\frac{(\alpha+1)}{(\beta-2)}+2\frac{\alpha}{(\beta-1)}\right\}$$

$$= \frac{\alpha(\alpha+1)(\alpha+2)}{(\beta-1)(\beta-2)(\beta-3)} + \frac{\alpha^2}{(\beta-2)(\beta-1)^3}\{-3(\alpha+1)(\beta-1)+2\alpha(\beta-2)\}$$

$$= \frac{\alpha(\alpha+1)(\alpha+2)}{(\beta-1)(\beta-2)(\beta-3)} + \frac{\alpha^2}{(\beta-2)(\beta-1)^3}\{-3(\alpha\beta+\beta-\alpha-1)+2(\alpha\beta-2\alpha)\}$$

$$= \frac{\alpha}{(\beta-2)(\beta-1)}\left[\frac{(\alpha+1)(\alpha+2)}{(\beta-3)} + \frac{\alpha(-\alpha\beta-\alpha-3\beta+3)}{(\beta-1)^2}\right]$$

$$= \frac{\alpha}{(\beta-1)^3(\beta-2)(\beta-3)}\left[(\alpha+1)(\alpha+2)(\beta-1)^2+\alpha(-\alpha\beta-\alpha-3\beta+3)(\beta-3)\right]$$

$$= \frac{\alpha}{(\beta-1)^3(\beta-2)(\beta-3)}$$

$$\left[(\alpha^2+3\alpha+2)(\beta^2+1-2\beta)+(-\alpha^2-\alpha^2\beta-3\alpha\beta+3\alpha)(\beta-3)\right]$$

$$= \frac{2\alpha}{(\beta-1)^3(\beta-2)(\beta-3)}\left[2\alpha^2-3\alpha+3\beta\alpha+\beta^2-2\beta+1\right]$$

$$= \frac{2\alpha}{(\beta-1)^3(\beta-2)(\beta-3)}\left[(2\alpha+\beta-1)(\alpha+\beta-1)\right] \qquad (4.4.8)$$

Similarly, we can find the fourth raw moment.

c. **Skewness and Kurtosis of Beta Distribution of Second Kind:** To find the shape of the distribution, we have derived the expressions of skewness (β_1, γ_1) and kurtosis (β_2, γ_2), which are given below:

$$\beta_1 = \frac{\mu_3^2}{\mu_2^3} = \frac{\left[\frac{2\alpha}{(\beta-1)^3(\beta-2)(\beta-3)}\left[(2\alpha+\beta-1)(\alpha+\beta-1)\right]\right]^2}{\left[\frac{\alpha(\beta+\alpha-1)}{(\beta-1)^2(\beta-2)}\right]^3}$$

$$\beta_1 = \frac{4(\beta-2)(2\alpha+\beta-1)^2}{\alpha(\beta-3)^2(\beta+\alpha-1)} \quad \text{[using Eqns. (4.4.7) and (4.4.8)]}$$

$$\gamma_1 = \sqrt{\beta_1} = \sqrt{\frac{4(\beta-2)(2\alpha+\beta-1)^2}{\alpha(\beta-3)^2(\beta+\alpha-1)}} = \frac{2(2\alpha+\beta-1)}{(\beta-3)}\sqrt{\frac{(\beta-2)}{\alpha(\beta+\alpha-1)}}.$$

Similarly, we can find the measure of kurtosis.

4.6 NORMAL DISTRIBUTION

A normal distribution is defined on the set of all real numbers $\mathbb{R}$. If a random variable X follows a normal distribution, the expression of the PDF is

$$f(x;\mu,\sigma^2) = \begin{cases} \frac{1}{\sqrt{2\pi}\sigma} e^{-\frac{1}{2}\left(\frac{x-\mu}{\sigma}\right)^2} & -\infty < x < \infty \\ 0 & \text{otherwise} \end{cases}$$

Symbolically: $X \sim N(\mu, \sigma^2)$

Here μ and σ^2 are the parameters of the distribution and the range of the parameter should be $-\infty < \mu < \infty$ and $\sigma^2 > 0$.

The CDF of the continuous normal distribution is (Figure 4.5):

$$F(x) = P(X \le x) = \int_{-\infty}^{x} f(t)dt = \int_{-\infty}^{x} \frac{1}{\sqrt{2\pi}\sigma} e^{-\frac{1}{2}\left(\frac{t-\mu}{\sigma}\right)^2} dt$$

$$F(x) = \int_{-\infty}^{x} \frac{1}{\sqrt{2\pi}\sigma} e^{-\frac{1}{2}\left(\frac{t-\mu}{\sigma}\right)^2} dt \quad -\infty < x < \infty$$

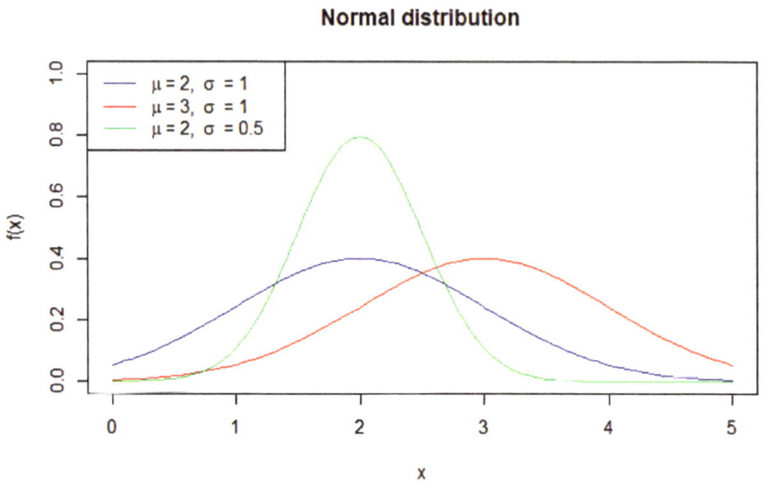

FIGURE 4.5 Normal distribution.

➢ Sample Generation of Normal Distribution

R Code: Generate a random sample from normal distribution.

x = rnorm(n, μ, σ)

where; n is the sample size; μ, σ are the parameters of the distribution

Example:

x = rnorm(10, 0, 1)

Output:

```
Console   Terminal   Jobs
> x=rnorm(10, 0, 1)
> x
 [1] -0.29895440  0.17485715 -0.07483009 -0.61241935 -0.67923481  1.53111726  0.17469421  0.49550681
 [9] -0.05968082 -1.33541726
>
```

4.6.1 PROPERTIES OF NORMAL DISTRIBUTION

a. **Moments of Normal Distribution:** The first four raw moments μ_1', μ_2', μ_3', μ_4' have been derived as below:

$$\mu_1' = E(X) = \int_{-\infty}^{\infty} x f(x) dx = \int_{-\infty}^{\infty} x \frac{1}{\sqrt{2\pi}\sigma} e^{-\frac{1}{2}\left(\frac{x-\mu}{\sigma}\right)^2} dx$$

$$Let\ \ u = \frac{x-\mu}{\sigma} \Rightarrow x = u\sigma + \mu$$

$$\Rightarrow du = \frac{1}{\sigma} dx \Rightarrow \quad dx = \sigma du$$

$$= \int_{-\infty}^{\infty} (u\sigma + \mu) \frac{1}{\sqrt{2\pi}\sigma} e^{-\frac{1}{2}u^2} \sigma du = \sigma \int_{-\infty}^{\infty} u \frac{1}{\sqrt{2\pi}} e^{-\frac{1}{2}u^2} du + \mu \int_{-\infty}^{\infty} \frac{1}{\sqrt{2\pi}} e^{-\frac{1}{2}u^2} du$$

$$\to \sigma \int_{-\infty}^{\infty} u \frac{1}{\sqrt{2\pi}} e^{-\frac{1}{2}u^2} du = 0 \qquad \text{[since } \int_{-\infty}^{\infty} u \frac{1}{\sqrt{2\pi}} e^{-\frac{1}{2}u^2} du \text{ is odd function]}$$

and $\int_{-\infty}^{\infty} \frac{1}{\sqrt{2\pi}} e^{-\frac{1}{2}u^2} du = 2\int_{0}^{\infty} \frac{1}{\sqrt{2\pi}} e^{-\frac{1}{2}u^2} du$ [since $\int_{-\infty}^{\infty} \frac{1}{\sqrt{2\pi}} e^{-\frac{1}{2}u^2} du$ is even function]

$$\Rightarrow 2\mu \frac{1}{\sqrt{2\pi}} \int_{0}^{\infty} e^{-\frac{1}{2}u^2} du$$

$$Let\ \ z = \frac{1}{2}u^2 \Rightarrow u = \sqrt{2z}$$

$$\Rightarrow dz = udu \quad \Rightarrow du = \frac{1}{\sqrt{2z}} dz$$

$$= \mu \frac{2}{\sqrt{2\pi}} \int_{0}^{\infty} e^{-z} \frac{1}{\sqrt{2z}} dz = \mu \frac{1}{\sqrt{\pi}} \int_{0}^{\infty} z^{\frac{1}{2}-1} e^{-z} dz = \mu \frac{1}{\sqrt{\pi}} \overline{|1/2} = \mu \frac{1}{\sqrt{\pi}} \sqrt{\pi}$$

$$= \mu \qquad (4.5.1)$$

Thus, *mean* of the normal distribution is μ.

$$\mu_2' = E(X^2) = \int_{-\infty}^{\infty} x^2 f(x) dx = \int_{-\infty}^{\infty} x^2 \frac{1}{\sqrt{2\pi}\sigma} e^{-\frac{1}{2}\left(\frac{x-\mu}{\sigma}\right)^2} dx$$

$$Let\ \ u = \frac{x-\mu}{\sigma} \Rightarrow x = u\sigma + \mu$$

$$\Rightarrow du = \frac{1}{\sigma} dx \Rightarrow \quad dx = \sigma du$$

$$= \int_{-\infty}^{\infty} (u\sigma + \mu)^2 \frac{1}{\sqrt{2\pi}\sigma} e^{-\frac{1}{2}u^2} \sigma du = \frac{1}{\sqrt{2\pi}} \int_{-\infty}^{\infty} (u\sigma + \mu)^2 e^{-\frac{1}{2}u^2} du$$

$$= \frac{1}{\sqrt{2\pi}} \int_{-\infty}^{\infty} \left\{(u\sigma)^2 + \mu^2 + 2\mu u\sigma)\right\} e^{-\frac{1}{2}u^2} du$$

$$= \frac{1}{\sqrt{2\pi}}\left\{\sigma^2 \int_{-\infty}^{\infty} u^2\, e^{-\frac{1}{2}u^2}\, du + \mu^2 \int_{-\infty}^{\infty} e^{-\frac{1}{2}u^2}\, du + 2\sigma\mu \int_{-\infty}^{\infty} u\, e^{-\frac{1}{2}u^2}\, du\right\}$$

$$\rightarrow 2\mu\sigma \int_{-\infty}^{\infty} u \frac{1}{\sqrt{2\pi}}\, e^{-\frac{1}{2}u^2}\, du = 0 \qquad \text{[odd function]}$$

and

$$\int_{-\infty}^{\infty} \frac{1}{\sqrt{2\pi}}\, e^{-\frac{1}{2}u^2}\, du = 2\int_{0}^{\infty} \frac{1}{\sqrt{2\pi}}\, e^{-\frac{1}{2}u^2}\, du \;\;\&\; \int_{-\infty}^{\infty} u^2 \frac{1}{\sqrt{2\pi}}\, e^{-\frac{1}{2}u^2}\, du = 2\int_{0}^{\infty} u^2 \frac{1}{\sqrt{2\pi}}\, e^{-\frac{1}{2}u^2}\, du$$

[even functions]

$$\Rightarrow \frac{2}{\sqrt{2\pi}}\left\{\sigma^2 \int_{0}^{\infty} u^2\, e^{-\frac{1}{2}u^2}\, du + \mu^2 \int_{0}^{\infty} e^{-\frac{1}{2}u^2}\, du\right\}$$

$$Let\;\; z = \frac{1}{2}u^2 \Rightarrow u = \sqrt{2z}$$

$$\Rightarrow dz = udu \quad \Rightarrow du = \frac{1}{\sqrt{2z}} dz$$

$$\Rightarrow \frac{2}{\sqrt{2\pi}}\left\{\sigma^2 \int_{0}^{\infty} 2z\, e^{-z} \frac{1}{\sqrt{2z}} dz + \mu^2 \int_{0}^{\infty} e^{-z} \frac{1}{\sqrt{2z}} dz\right\}$$

$$= \frac{1}{\sqrt{\pi}}\left\{2\sigma^2 \int_{0}^{\infty} z^{\frac{3}{2}-1}\, e^{-z} dz + \mu^2 \int_{0}^{\infty} z^{\frac{1}{2}-1}\, e^{-z} dz\right\}$$

$$= \frac{1}{\sqrt{\pi}}\left\{2\sigma^2\Gamma(3/2) + \mu^2\Gamma(1/2)\right\} = \frac{1}{\sqrt{\pi}}\left\{2\sigma^2 \frac{1}{2}\Gamma(1/2) + \mu^2\Gamma(1/2)\right\}$$

$$= \frac{1}{\sqrt{\pi}}\left\{\sigma^2\sqrt{\pi} + \mu^2\sqrt{\pi}\right\}$$

$$\mu^2 + \sigma^2 \qquad (4.5.2)$$

$$\mu_3' = E(X^3) = \int_{-\infty}^{\infty} x^3 f(x) dx = \int_{-\infty}^{\infty} x^3 \frac{1}{\sqrt{2\pi}\sigma}\, e^{-\frac{1}{2}\left(\frac{x-\mu}{\sigma}\right)^2} dx$$

$$Let\;\; u = \frac{x-\mu}{\sigma} \Rightarrow x = u\sigma + \mu$$

$$\Rightarrow du = \frac{1}{\sigma} dx \;\Rightarrow\quad dx = \sigma du$$

$$= \int_{-\infty}^{\infty} (u\sigma+\mu)^3 \frac{1}{\sqrt{2\pi}\sigma} e^{-\frac{1}{2}u^2} \sigma du = \frac{1}{\sqrt{2\pi}} \int_{-\infty}^{\infty} (u\sigma+\mu)^3 e^{-\frac{1}{2}u^2} du$$

$$= \frac{1}{\sqrt{2\pi}} \int_{-\infty}^{\infty} \left\{ (u\sigma)^3 + \mu^3 + 3(u\sigma)^2 \mu + 3\mu^2 u\sigma \right\} e^{-\frac{1}{2}u^2} du$$

$$= \frac{1}{\sqrt{2\pi}} \left\{ \sigma^3 \int_{-\infty}^{\infty} u^3 e^{-\frac{1}{2}u^2} du + \mu^3 \int_{-\infty}^{\infty} e^{-\frac{1}{2}u^2} du + 3\sigma\mu^2 \int_{-\infty}^{\infty} u e^{-\frac{1}{2}u^2} du + 3\sigma^2\mu \int_{-\infty}^{\infty} u^2 e^{-\frac{1}{2}u^2} du \right\}$$

$$\sigma^3 \int_{-\infty}^{\infty} u^3 e^{-\frac{1}{2}u^2} du = 3\sigma\mu^2 \int_{-\infty}^{\infty} u e^{-\frac{1}{2}u^2} du = 0 \qquad \text{[odd functions]}$$

and

$$\mu^3 \int_{-\infty}^{\infty} e^{-\frac{1}{2}u^2} du = 2\mu^3 \int_{0}^{\infty} e^{-\frac{1}{2}u^2} du \;\; \& \;\; 3\sigma^2\mu \int_{-\infty}^{\infty} u^2 e^{-\frac{1}{2}u^2} du = 6\sigma^2\mu \int_{0}^{\infty} u^2 e^{-\frac{1}{2}u^2} du$$

[even functions]

$$Let \;\; z = \frac{1}{2}u^2 \Rightarrow u = \sqrt{2z}$$

$$\Rightarrow dz = udu \quad \Rightarrow du = \frac{1}{\sqrt{2z}} dz$$

$$= \frac{2}{\sqrt{2\pi}} \left\{ \mu^3 \int_{0}^{\infty} e^{-z} \frac{1}{\sqrt{2z}} dz + 3\sigma^2\mu \int_{0}^{\infty} 2z\, e^{-z} \frac{1}{\sqrt{2z}} dz \right\}$$

$$= \frac{1}{\sqrt{\pi}} \left\{ \mu^3 \int_{0}^{\infty} z^{\frac{1}{2}-1} e^{-z} dz + 6\sigma^2\mu \int_{0}^{\infty} z^{\frac{3}{2}-1} e^{-z} dz \right\}$$

$$= \frac{1}{\sqrt{\pi}} \left\{ \mu^3 \Gamma(1/2) + 6\sigma^2\mu\, \Gamma(3/2) \right\} = \frac{1}{\sqrt{\pi}} \left\{ \mu^3 \sqrt{\pi} + 6\sigma^2\mu \frac{1}{2}\sqrt{\pi} \right\}$$

$$= \mu^3 + 3\sigma^2\mu \qquad (4.5.3)$$

$$\mu_4' = E(X^4) = \int_{-\infty}^{\infty} x^4 f(x) dx = \int_{-\infty}^{\infty} x^4 \frac{1}{\sqrt{2\pi}\sigma} e^{-\frac{1}{2}\left(\frac{x-\mu}{\sigma}\right)^2} dx$$

$$Let \;\; u = \frac{x-\mu}{\sigma} \Rightarrow x = u\sigma + \mu$$

$$\Rightarrow du = \frac{1}{\sigma} dx \;\Rightarrow\; dx = \sigma du$$

$$= \int_{-\infty}^{\infty} (u\sigma+\mu)^4 \frac{1}{\sqrt{2\pi}\sigma} e^{-\frac{1}{2}u^2} \sigma du = \frac{1}{\sqrt{2\pi}} \int_{-\infty}^{\infty} (u\sigma+\mu)^4 e^{-\frac{1}{2}u^2} du$$

$$= \frac{1}{\sqrt{2\pi}} \int_{-\infty}^{\infty} \left\{ (u\sigma)^4 + \mu^4 + 4(u\sigma)^3 \mu + 4\mu^3 u\sigma + 6(u\sigma)^2 \mu^2 \right\} e^{-\frac{1}{2}u^2} du$$

$$= \frac{1}{\sqrt{2\pi}} \left\{ \begin{array}{l} \sigma^4 \int_{-\infty}^{\infty} u^4 e^{-\frac{1}{2}u^2} du + \mu^4 \int_{-\infty}^{\infty} e^{-\frac{1}{2}u^2} du + 4\sigma\mu^3 \int_{-\infty}^{\infty} u e^{-\frac{1}{2}u^2} du + 4\sigma^3 \mu \\ \int_{-\infty}^{\infty} u^3 e^{-\frac{1}{2}u^2} du + 6\sigma^2\mu^2 \int_{-\infty}^{\infty} u^2 e^{-\frac{1}{2}u^2} du \end{array} \right\}$$

$$\to 4\sigma\mu^3 \int_{-\infty}^{\infty} u e^{-\frac{1}{2}u^2} du = 4\sigma^3 \mu \int_{-\infty}^{\infty} u^3 e^{-\frac{1}{2}u^2} du = 0 \quad \text{[odd functions]}$$

and

$$\to \left[\begin{array}{l} \sigma^4 \int_{-\infty}^{\infty} u^4 e^{-\frac{1}{2}u^2} du = 2\sigma^4 \int_0^{\infty} u^4 e^{-\frac{1}{2}u^2} du, \mu^4 \int_{-\infty}^{\infty} e^{-\frac{1}{2}u^2} du = 2\mu^4 \int_0^{\infty} e^{-\frac{1}{2}u^2} du \\ \& \ 6\sigma^2\mu^2 \int_{-\infty}^{\infty} u^2 e^{-\frac{1}{2}u^2} du = 12\sigma^2\mu^2 \int_0^{\infty} u^2 e^{-\frac{1}{2}u^2} du \end{array} \right]$$

[even functions]

$$\Rightarrow \frac{2}{\sqrt{2\pi}} \left\{ \sigma^4 \int_0^{\infty} u^4 e^{-\frac{1}{2}u^2} du + \mu^4 \int_0^{\infty} e^{-\frac{1}{2}u^2} du + 6\sigma^2\mu^2 \int_0^{\infty} u^2 e^{-\frac{1}{2}u^2} du \right\}$$

$$Let \ z = \frac{1}{2}u^2 \Rightarrow u = \sqrt{2z}$$

$$\Rightarrow dz = udu \quad \Rightarrow du = \frac{1}{\sqrt{2z}} dz$$

$$= \frac{2}{\sqrt{2\pi}} \left\{ \sigma^4 \int_0^{\infty} (2z)^2 e^{-z} \frac{1}{\sqrt{2z}} dz + \mu^4 \int_0^{\infty} e^{-z} \frac{1}{\sqrt{2z}} dz + 6\sigma^2\mu^2 \int_0^{\infty} 2z e^{-z} \frac{1}{\sqrt{2z}} dz \right\}$$

$$= \frac{1}{\sqrt{\pi}} \left\{ 4\sigma^4 \int_0^{\infty} z^2 e^{-z} \frac{1}{\sqrt{z}} dz + \mu^4 \int_0^{\infty} e^{-z} \frac{1}{\sqrt{z}} dz + 12\sigma^2\mu^2 \int_0^{\infty} z e^{-z} \frac{1}{\sqrt{z}} dz \right\}$$

$$= \frac{1}{\sqrt{\pi}} \left\{ 4\sigma^4 \int_0^{\infty} z^{\frac{5}{2}-1} e^{-z} \frac{1}{\sqrt{z}} dz + \mu^4 \int_0^{\infty} z^{\frac{1}{2}-1} e^{-z} dz + 12\sigma^2\mu^2 \int_0^{\infty} z^{\frac{3}{2}-1} e^{-z} dz \right\}$$

$$= \frac{1}{\sqrt{\pi}}\left\{4\sigma^4\Gamma(5/2) + \mu^4\Gamma(1/2) + 12\sigma^2\mu^2\Gamma(3/2)\right\}$$

$$= \frac{1}{\sqrt{\pi}}\left\{4\sigma^4\frac{3}{2}\frac{1}{2}\Gamma(1/2) + \mu^4\Gamma(1/2) + 12\sigma^2\mu^2\frac{1}{2}\Gamma(1/2)\right\}$$

$$= \frac{1}{\sqrt{\pi}}\left\{3\sigma^4\sqrt{\pi} + \mu^4\sqrt{\pi} + 6\sigma^2\mu^2\sqrt{\pi}\right\}$$

$$= 3\sigma^4 + \mu^4 + 6\sigma^2\mu^2 \qquad (4.5.4)$$

b. **Central Moments of Normal Distribution:** The first four central moments μ_1, μ_2, μ_3, μ_4 have been derived as below:

$$\mu = 0 \text{ [first central moments always zero]} \qquad (4.5.5)$$

$$\mu_2 = \mu_2' - \mu_1'^2$$

$$\mu_2 = \sigma^2 \text{ [using Eqns. (4.5.1) and (4.5.2)]} \qquad (4.5.6)$$

Thus, *variance* of the uniform discrete distribution is σ^2

$$\mu_3 = \mu_3' - 3\mu_2'\mu_1' + 2\mu_1'^3$$

$$\mu_3 = 0 \text{ [using Eqns. (4.5.1)–(4.5.3)]} \qquad (4.5.7)$$

$$\mu_4 = \mu_4' - 4\mu_3'\mu_1' + 6\mu_2'\mu_1'^2 - 3\mu_1'^4$$

$$\mu_4 = 3\sigma^4 \text{ [using Eqns. (4.5.1)–(4.5.4)]} \qquad (4.5.8)$$

c. **Skewness and Kurtosis of Normal Distribution:** To find the shape of the distribution, we have derived the expressions of skewness (β_1, γ_1) and kurtosis (β_2, γ_2), which are given below:

$$\beta_1 = \frac{\mu_3^2}{\mu_2^3} \text{ [using Eqns. (4.5.6) and (4.5.7)]}$$

$$\beta_1 = 0$$

$$\gamma_1 = \sqrt{\beta_1} = 0$$

$$\beta_2 = \frac{\mu_4}{\mu_2^2} = 3 \text{ [using Eqns. (4.5.6) and (4.5.8)]}$$

$$\gamma_2 = \beta_2 - 3 = 0$$

This implies that the distribution is symmetric and mesokurtic.

d. Moment Generating Function (MGF) of Normal Distribution: The expression for MGF ($M_X(t)$) is obtained as below:

$$M_X(t) = E(e^{tX}) = \int_{-\infty}^{\infty} e^{tx} f(x)dx = \int_{-\infty}^{\infty} e^{tx} \frac{1}{\sqrt{2\pi}\sigma} e^{-\frac{1}{2}\left(\frac{x-\mu}{\sigma}\right)^2} dx$$

$$Let\ \ u = \frac{x-\mu}{\sigma} \Rightarrow x = u\sigma + \mu$$

$$\Rightarrow du = \frac{1}{\sigma} dx \Rightarrow \quad dx = \sigma du$$

$$= \int_{-\infty}^{\infty} e^{t(u\sigma+\mu)} \frac{1}{\sqrt{2\pi}\sigma} e^{-\frac{1}{2}u^2} \sigma du = \frac{1}{\sqrt{2\pi}} e^{t\mu} \int_{-\infty}^{\infty} e^{tu\sigma} e^{-\frac{1}{2}u^2} du = \frac{1}{\sqrt{2\pi}} e^{t\mu} \int_{-\infty}^{\infty} e^{-\frac{1}{2}\left(u^2-2tu\sigma\right)} du$$

$$= \frac{1}{\sqrt{2\pi}} e^{t\mu} \int_{-\infty}^{\infty} e^{-\frac{1}{2}\left(u^2-2tu\sigma+t^2\sigma^2-t^2\sigma^2\right)} du = \frac{1}{\sqrt{2\pi}} e^{t\mu+\frac{1}{2}t^2\sigma^2} \int_{-\infty}^{\infty} e^{-\frac{1}{2}(u-t\sigma)^2} du$$

$$Let\ \ s = u - t\sigma \Rightarrow u = s + t\sigma$$

$$\Rightarrow ds = du$$

$$= e^{t\mu+\frac{1}{2}t^2\sigma^2} \int_{-\infty}^{\infty} \frac{1}{\sqrt{2\pi}} e^{-\frac{1}{2}s^2} ds$$

$$Let\ \ p = \frac{s^2}{2} \Rightarrow 2dp = 2sds \Rightarrow ds = \frac{1}{s} dp \Rightarrow ds = \frac{1}{\sqrt{2p}} dp$$

$$= e^{t\mu+\frac{1}{2}t^2\sigma^2} \int_{-\infty}^{\infty} \frac{1}{\sqrt{2\pi}} e^{-p} \frac{1}{\sqrt{2p}} dp = 2e^{t\mu+\frac{1}{2}t^2\sigma^2} \int_{0}^{\infty} \frac{1}{2\sqrt{\pi}} p^{-1/2} e^{-p} dp = e^{t\mu+\frac{1}{2}t^2\sigma^2} \int_{0}^{\infty} \frac{1}{\sqrt{\pi}} p^{1/2-1} e^{-p} dp,$$

Using gamma function, we get:

$$= \frac{1}{\sqrt{\pi}} e^{t\mu+\frac{1}{2}t^2\sigma^2} \Gamma(1/2) = \frac{1}{\sqrt{\pi}} e^{t\mu+\frac{1}{2}t^2\sigma^2} \sqrt{\pi}$$

$$\Rightarrow M_X(t) = e^{t\mu+\frac{1}{2}t^2\sigma^2}$$

Note: The expression PGF can be derived using the same MGF method. The formula of PGF is $P_X(t) = \int_x t^x f(x)dx.$

e. Characteristic Function (CF) of Normal Distribution: The expression for CF ($\phi_X(t)$) is obtained as below:

$$\phi_X(t) = E\left(e^{itx}\right) = \int_{-\infty}^{\infty} e^{itx} f(x)dx = \int_{-\infty}^{\infty} e^{itx} \frac{1}{\sqrt{2\pi}\sigma} e^{-\frac{1}{2}\left(\frac{x-\mu}{\sigma}\right)^2} dx.$$

Note that this integral is a complex-valued integral, which is beyond the scope of this book. Therefore, we will state the characteristic function without proof.

$$\Rightarrow \phi_X(t) = e^{it\mu - \frac{1}{2}t^2\sigma^2}$$

f. Cumulants Generating Function of Normal Distribution: The expression for CGF ($K_X(t)$) is obtained as below:

$$K_X(t) = \log(M_X(t))$$

$$K_X(t) = \log\left(e^{t\mu + \frac{1}{2}t^2\sigma^2}\right) = t\mu + \frac{1}{2}t^2\sigma^2$$

We can get central moments using cumulants generating function:

$$\kappa_1 = \text{mean}; \quad \kappa_2 = \text{variance}; \quad \kappa_3 = \mu_3; \quad \kappa_4 = -3\kappa_2 + \mu_4.$$

$$K_X(t) = t\mu + \frac{1}{2}t^2\sigma^2.$$

$$\kappa_r = \text{coefficient of } \frac{t^r}{r!} \text{ in } K_X(t).$$

$$\kappa_1 = \text{mean} = \text{coefficient of } \frac{t}{1!} \text{ in } K_X(t) = \mu.$$

$$\kappa_2 = \text{variance} = \text{coefficient of } \frac{t^2}{2!} \text{ in } K_X(t) = \sigma^2.$$

$$\kappa_3 = \mu_3 = \text{coefficient of } \frac{t^3}{3!} \text{ in } K_X(t) = 0.$$

$$\kappa_4 = \text{coefficient of } \frac{t^4}{4!} \text{ in } K_X(t) = 0.$$

$$\mu_4 = \kappa_4 + 3\kappa_2^2 = 0 + 3\sigma^2 = 3\sigma^2.$$

g. General Central Moment of Normal Distribution: Case I: Odd order central moment of normal distribution

$$\mu_{2n+1} = E(X-\mu)^{2n+1} = \int_{-\infty}^{\infty} (x-\mu)^{2n+1} f(x)dx = \int_{-\infty}^{\infty} (x-\mu)^{2n+1} \frac{1}{\sqrt{2\pi}\sigma} e^{-\frac{1}{2}\left(\frac{x-\mu}{\sigma}\right)^2} dx$$

$$Let\ \ u = \frac{x-\mu}{\sigma} \Rightarrow x = u\sigma + \mu$$

$$\Rightarrow du = \frac{1}{\sigma}dx \Rightarrow \quad dx = \sigma du$$

$$= \int_{-\infty}^{\infty} (u\sigma)^{2n+1} \frac{1}{\sqrt{2\pi}\sigma} e^{-\frac{1}{2}u^2} \sigma du = \frac{1}{\sqrt{2\pi}} \sigma^{(2n+1)} \int_{-\infty}^{\infty} u^{(2n+1)} e^{-\frac{1}{2}u^2} du$$

[Since the above function is odd]

$$\mu_{2n+1} = 0$$

So, all the odd central moments are 0 ($\mu_1 = \mu_3 = \mu_5 = \ldots = \mu_{2n+1} = 0$)

Case II: Even order central moment of normal distribution

$$\mu_{2n} = E(X-\mu)^{2n} = \int_{-\infty}^{\infty} (x-\mu)^{2n} f(x)dx = \int_{-\infty}^{\infty} (x-\mu)^{2n} \frac{1}{\sqrt{2\pi}\sigma} e^{-\frac{1}{2}\left(\frac{x-\mu}{\sigma}\right)^2} dx$$

$$Let\ \ u = \frac{x-\mu}{\sigma} \Rightarrow x = u\sigma + \mu$$

$$\Rightarrow du = \frac{1}{\sigma}dx \Rightarrow \quad dx = \sigma du$$

$$= \int_{-\infty}^{\infty} (u\sigma)^{2n} \frac{1}{\sqrt{2\pi}\sigma} e^{-\frac{1}{2}u^2} \sigma du = \frac{1}{\sqrt{2\pi}} \sigma^{(2n)} \int_{-\infty}^{\infty} u^{(2n)} e^{-\frac{1}{2}u^2} du$$

$$Let\ \ p = \frac{u^2}{2} \Rightarrow 2dp = 2udu \Rightarrow du = \frac{1}{\sqrt{2p}} dp$$

$$= \frac{1}{\sqrt{2\pi}} \sigma^{(2n)} \int_{-\infty}^{\infty} (2p)^{(n)} e^{-p} \frac{1}{\sqrt{2p}} dp = \sigma^{(2n)} (2)^{n+1}$$

$$\int_{0}^{\infty} \frac{1}{2\sqrt{\pi}} p^{n-1/2} e^{-p} dp = (2)^n \sigma^{(2n)} \int_{0}^{\infty} \frac{1}{\sqrt{\pi}} p^{n+1/2-1} e^{-p} dp$$

$$\mu_{2n} = \frac{2^n \sigma^{2n} \left(n+\frac{1}{2}\right)!}{\sqrt{\pi}}$$

Case n=1

$$\mu_2 = \frac{2\sigma^2 \left(1+\frac{1}{2}\right)!}{\sqrt{\pi}} = \frac{2\sigma^2 \left(\frac{3}{2}\right)!}{\sqrt{\pi}} = \sigma^2$$

Case n=2

$$\mu_4 = \frac{2^2\sigma^4\left(2+\frac{1}{2}\right)!}{\sqrt{\pi}} = \frac{4\sigma^4\left(\frac{5}{2}\right)!}{\sqrt{\pi}} = 3\sigma^2$$

Similarly, we get $\mu_6 = \mu_8, \ldots, \mu_{2n}$

h. Mode of Normal Distribution: Mode is the value of x for which $f(x)$ is maximum, i.e., the mode is the solution of $f'(x) = 0$ and

$$f''(x) = 0$$

We have:

$$f(x;\mu,\sigma^2) = \begin{cases} \frac{1}{\sqrt{2\pi}\sigma} e^{-\frac{1}{2}\left(\frac{x-\mu}{\sigma}\right)^2} & -\infty < x < \infty \\ 0 & \text{otherwise} \end{cases}$$

$$\log f(x) = \log\left\{\frac{1}{\sqrt{2\pi}\sigma} e^{-\frac{1}{2}\left(\frac{x-\mu}{\sigma}\right)^2}\right\} = \log\left\{\frac{1}{\sqrt{2\pi}\sigma}\right\} + \log\left\{e^{-\frac{1}{2}\left(\frac{x-\mu}{\sigma}\right)^2}\right\}$$

$$= \log\left\{\frac{1}{\sqrt{2\pi}\sigma}\right\} - \frac{1}{2}\left(\frac{x-\mu}{\sigma}\right)^2$$

$$\frac{\partial}{\partial x}\log f(x) = \frac{\partial}{\partial x}\left[\log\left\{\frac{1}{\sqrt{2\pi}\sigma}\right\} - \frac{1}{2}\left(\frac{x-\mu}{\sigma}\right)^2\right] = -\left(\frac{x-\mu}{\sigma^2}\right)$$

$$\frac{1}{f(x)}f'(x) = -\left(\frac{x-\mu}{\sigma^2}\right) \Rightarrow f'(x) = -\left(\frac{x-\mu}{\sigma^2}\right)f(x)$$

$$\left(\frac{x-\mu}{\sigma^2}\right)f(x) = 0$$

$$\Rightarrow x-\mu = 0 \Rightarrow x = \mu \qquad [\textit{Since } f(x) \ \& \ \sigma^2 > 0]$$

Now,

$$\frac{\partial}{\partial x}f'(x) = \frac{\partial}{\partial x}\left[-\left(\frac{x-\mu}{\sigma^2}\right)f(x)\right] = -\frac{1}{\sigma^2}\left[\frac{\partial}{\partial x}(x-\mu)f(x)\right]$$

$$= -\frac{1}{\sigma^2}\left[f(x)+(x-\mu)f'(x)\right]$$

Substitute the value of $f'(x)$ in the above function:

$$\frac{\partial}{\partial x} f'(x) = -\frac{1}{\sigma^2}\left[f(x) - \frac{(x-\mu)^2}{\sigma^2} f(x) \right] = -\frac{f(x)}{\sigma^2}\left[1 - \frac{(x-\mu)^2}{\sigma^2} \right]$$

$$f''(x) = -\left\{ \frac{f(x)}{\sigma^2}\left[1 - \frac{(x-\mu)^2}{\sigma^2} \right] \right\} = \left(-\frac{f(x)}{\sigma^2} \right)_{x=\mu} < 0.$$

Hence,
$x = \mu$, is the mode of the normal distribution.

4.7 LOG-NORMAL DISTRIBUTION

A log-normal distribution is defined on the interval $(0, \infty)$. If $\log_e X$ follows a normal distribution, then a positive random variable X follows a log-normal distribution. In other words, if $\log_e X \sim N(\mu, \sigma^2)$ then X ~ Log-normal distribution.

The expression of the PDF of random variable X is:

$$f(x;\mu,\sigma^2) = \begin{cases} \dfrac{1}{x\sqrt{2\pi}\sigma} e^{-\frac{1}{2}\left(\frac{\log_e x-\mu}{\sigma}\right)^2} & \text{if } 0 < x < \infty \\ 0 & \text{otherwise} \end{cases}$$

Here μ and σ^2 are two parameters of the distribution and the range of the parameters should be $\mu \in \mathbb{R}$ and $\sigma > 0$.

Remark: If $\log_e X$ follows a standard normal distribution, then a positive random variable X follows standard log-normal distribution (Figure 4.6).

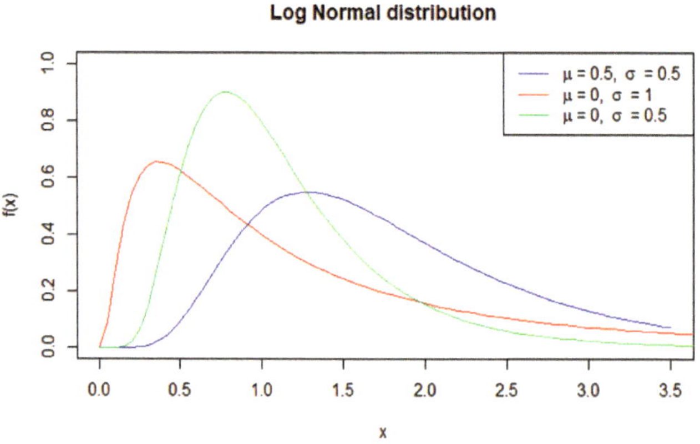

FIGURE 4.6 Log normal distribution.

➢ **Sample Generation of Log-Normal Distribution**

R Code: Generate a random sample from Log-Normal distribution.

x = rlnorm(n, μ, σ)

where; n is the sample size; μ, σ are the parameters of the distribution

Example:

x = rlnorm(10, 0.5, 1)

Output:

```
Console  Terminal  Jobs
> x=rlnorm(10, 0.5, 1)
> x
 [1] 0.4893407 4.6660692 4.3236624 3.1572318 0.7106875 1.0981364 3.2375026 0.6828865 0.5962930 1.6046959
>
```

4.7.1 *PROPERTIES OF THE LOG-NORMAL DISTRIBUTION*

a. **Distribution Function of Log Normal Distribution:**

$$F_X(x) = P\left(X \le x\right) = P\left(\log_e X \le \log_e x\right)$$

$$= P\left(\frac{\log_e X - \mu}{\sigma} \le \frac{\log_e x - \mu}{\sigma}\right) = P\left(Z \le \frac{\log_e x - \mu}{\sigma}\right),$$

where; $Z \sim N(0,1)$

$$F_X(x) = \Phi\left(\frac{\log_e x - \mu}{\sigma}\right),$$

where; Φ denotes the CDF of N(0,1)

b. **Median of Log Normal Distribution:** Let X~Log normal distribution with parameters μ and σ^2 and $\log_e X \sim N(\mu\ \sigma^2)$. Then

Median($\log_e$ X)= μ

$\log_e$(Median X)= μ

[If f(x) is a monotonic function. Then median of f(x) = f(median(x))]

$\Rightarrow$ Median = e^{μ}

So, the median of the log-normal distribution is e^{μ}.

c. **Mode of Log Normal Distribution:**

The PDF of the log-normal distribution is given below:

$$f(x)=\begin{cases}\dfrac{1}{x\sqrt{2\pi}\sigma}\, e^{-\frac{1}{2}\left(\frac{\log_e x-\mu}{\sigma}\right)^2} & \text{if } 0<x<\infty \\ 0 & \text{otherwise}\end{cases}$$

For x > 0,

$$\log_e f(x)=\log_e\left[\frac{1}{x\sqrt{2\pi}\sigma}\, e^{-\frac{1}{2}\left(\frac{\log_e x-\mu}{\sigma}\right)^2}\right]$$

$$\log_e f(x)=-\log_e x-\log_e\left(\sqrt{2\pi}\sigma\right)-\frac{1}{2}\left(\frac{\log_e x-\mu}{\sigma}\right)^2$$

$$\frac{d}{dx}\log_e f(x)=-\log_e x-\log_e\left(\sqrt{2\pi}\sigma\right)-\frac{1}{2}\left(\frac{\log_e x-\mu}{\sigma}\right)^2$$

$$\frac{f'(x)}{f(x)}=-\frac{1}{x}-\frac{1}{2}2\left(\frac{\log_e x-\mu}{\sigma}\right)\frac{1}{\sigma x}$$

$$f'(x)=-\frac{f(x)}{x}\left[1+\left(\frac{\log_e x-\mu}{\sigma^2}\right)\right]$$

Note that $f'(x)=0$

$$\Rightarrow -\frac{f(x)}{x}\left[1+\left(\frac{\log_e x-\mu}{\sigma^2}\right)\right]=0$$

As we know that $f(x)$ & $x\neq 0$

$$\Rightarrow 1+\left(\frac{\log_e x-\mu}{\sigma^2}\right)=0$$

$$\Rightarrow \frac{\log_e x-\mu}{\sigma^2}=-1$$

$$\Rightarrow \log_e x-\mu=-\sigma^2$$

$$\Rightarrow x=e^{\mu-\sigma^2}$$

$$f'(x)=\begin{cases}>0 & x<e^{\mu-\sigma^2} \\ <0 & x>e^{\mu-\sigma^2}\end{cases}$$

By 1st derivative test, f(x) is maximum at $x=e^{\mu-\sigma^2}$

Hence $x = e^{\mu-\sigma^2}$ is the mode of the distribution.

Remark:

Mean $= e^{\mu+\frac{1}{2}\sigma^2}$ > Median $= e^{\mu}$ > Mode $= e^{\mu-\sigma^2}$

The derivation of the mean is derived in the next coming section of "Moments of Log Normal Distribution."

Hence, Log-normal Distribution is positively skewed.

d. Moments of Log Normal Distribution: The first four raw moments $\mu_1', \mu_2', \mu_3', \mu_4'$ have been derived as below:

$$\mu_1' = E(X) = \int_0^\infty x f(x)dx = \int_0^\infty x \frac{1}{x\sqrt{2\pi}\sigma} e^{-\frac{1}{2}\left(\frac{\log_e x-\mu}{\sigma}\right)^2} dx$$

$$Let\ \ u = \frac{\log_e x - \mu}{\sigma} \Rightarrow x = e^{(u\sigma+\mu)}$$

$$\Rightarrow du = \frac{1}{x\sigma}dx \Rightarrow \quad dx = x\sigma du \ \ \Rightarrow dx = e^{(u\sigma+\mu)}\sigma du$$

Using the above results, we get:

$$= \int_{-\infty}^{\infty} e^{(u\sigma+\mu)} \frac{1}{\sqrt{2\pi}\sigma} e^{-\frac{1}{2}u^2} \sigma du = e^{\mu} \frac{1}{\sqrt{2\pi}} \int_{-\infty}^{\infty} e^{u\sigma} e^{-\frac{1}{2}u^2} du = e^{\mu} \frac{1}{\sqrt{2\pi}} \int_{-\infty}^{\infty} e^{-\frac{1}{2}\left(u^2-2u\sigma\right)} du$$

$$= e^{\mu} \frac{1}{\sqrt{2\pi}} \int_{-\infty}^{\infty} e^{-\frac{1}{2}(u-\sigma)^2+\frac{1}{2}\sigma^2} du = e^{\mu+\frac{1}{2}\sigma^2} \frac{1}{\sqrt{2\pi}} \int_{-\infty}^{\infty} e^{-\frac{1}{2}(u-\sigma)^2} du$$

$$Let\ \ p = u - \sigma \Rightarrow dp = du$$

Applying the above results, we get:

$$= e^{\mu+\frac{1}{2}\sigma^2} \int_{-\infty}^{\infty} \frac{1}{\sqrt{2\pi}} e^{-\frac{1}{2}p^2} dp$$

As we know that $\int_{-\infty}^{\infty} \frac{1}{\sqrt{2\pi}} e^{-\frac{1}{2}p^2} dp = 1$, therefore:

$$= e^{\mu+\frac{1}{2}\sigma^2}. \tag{4.6.1}$$

So, *mean* of the log-normal distribution is $e^{\mu+\frac{1}{2}\sigma^2}$.

$$\mu_2' = E(X^2) = \int_0^\infty x^2 f(x)dx = \int_0^\infty x^2 \frac{1}{x\sqrt{2\pi}\sigma} e^{-\frac{1}{2}\left(\frac{\log_e x-\mu}{\sigma}\right)^2} dx$$

$$Let\ \ u = \frac{\log_e x - \mu}{\sigma} \Rightarrow x = e^{(u\sigma + \mu)}$$

$$\Rightarrow du = \frac{1}{x\sigma} dx \Rightarrow \quad dx = x\sigma du \quad \Rightarrow dx = e^{(u\sigma + \mu)} \sigma du$$

Substituting the above results, we get:

$$= \int_{-\infty}^{\infty} e^{2(u\sigma+\mu)} \frac{1}{\sqrt{2\pi}\sigma} e^{-\frac{1}{2}u^2} \sigma du = e^{2\mu} \frac{1}{\sqrt{2\pi}} \int_{-\infty}^{\infty} e^{2u\sigma} e^{-\frac{1}{2}u^2} du = e^{2\mu} \frac{1}{\sqrt{2\pi}} \int_{-\infty}^{\infty} e^{-\frac{1}{2}\left(u^2 - 4u\sigma\right)} du$$

$$Let\ \ p = u - 2\sigma \Rightarrow dp = du$$

Using the above results, we get:

$$= e^{2\mu + 2\sigma^2} \int_{-\infty}^{\infty} \frac{1}{\sqrt{2\pi}} e^{-\frac{1}{2}p^2} dp$$

As we know that $\int_{-\infty}^{\infty} \frac{1}{\sqrt{2\pi}} e^{-\frac{1}{2}p^2} dp = 1,$ therefore:

$$= e^{2\mu + 2\sigma^2} \tag{4.6.2}$$

$$\mu_3' = E(X^3) = \int_0^{\infty} x^3 f(x) dx = \int_0^{\infty} x^3 \frac{1}{x\sqrt{2\pi}\sigma} e^{-\frac{1}{2}\left(\frac{\log_e x - \mu}{\sigma}\right)^2} dx$$

$$Let\ \ u = \frac{\log_e x - \mu}{\sigma} \Rightarrow x = e^{(u\sigma + \mu)}$$

$$\Rightarrow du = \frac{1}{x\sigma} dx \Rightarrow \quad dx = x\sigma du \quad \Rightarrow dx = e^{(u\sigma + \mu)} \sigma du$$

Using the above results, we get:

$$= \int_{-\infty}^{\infty} e^{3(u\sigma+\mu)} \frac{1}{\sqrt{2\pi}\sigma} e^{-\frac{1}{2}u^2} \sigma du = e^{3\mu} \frac{1}{\sqrt{2\pi}}$$

$$\int_{-\infty}^{\infty} e^{3u\sigma} e^{-\frac{1}{2}u^2} du = e^{3\mu} \frac{1}{\sqrt{2\pi}} \int_{-\infty}^{\infty} e^{-\frac{1}{2}\left(u^2 - 6u\sigma\right)} du$$

$$= e^{3\mu} \frac{1}{\sqrt{2\pi}} \int_{-\infty}^{\infty} e^{-\frac{1}{2}(u-3\sigma)^2 + \frac{9}{2}\sigma^2} du = e^{3\mu + \frac{9}{2}\sigma^2} \frac{1}{\sqrt{2\pi}} \int_{-\infty}^{\infty} e^{-\frac{1}{2}(u-3\sigma)^2} du$$

$$Let\ \ p = u - 3\sigma \Rightarrow dp = du$$

Applying the results mentioned above, we have:

$$= e^{3\mu+\frac{9}{2}\sigma^2}\int_{-\infty}^{\infty}\frac{1}{\sqrt{2\pi}}e^{-\frac{1}{2}p^2}dp$$

As we know that $\int_{-\infty}^{\infty}\frac{1}{\sqrt{2\pi}}e^{-\frac{1}{2}p^2}dp = 1,$ therefore:

$$= e^{3\mu+\frac{9}{2}\sigma^2}. \tag{4.6.3}$$

$$\mu_4' = E(X^4) = \int_0^{\infty} x^4 f(x)dx = \int_0^{\infty} x^4 \frac{1}{x\sqrt{2\pi}\sigma}e^{-\frac{1}{2}\left(\frac{\log_e x-\mu}{\sigma}\right)^2}dx$$

$$Let\ \ u = \frac{\log_e x - \mu}{\sigma} \Rightarrow x = e^{(u\sigma+\mu)}$$

$$\Rightarrow du = \frac{1}{x\sigma}dx \Rightarrow \quad dx = x\sigma du \ \Rightarrow dx = e^{(u\sigma+\mu)}\sigma du$$

Using the above results, we get:

$$= \int_{-\infty}^{\infty} e^{4(u\sigma+\mu)}\frac{1}{\sqrt{2\pi}\sigma}e^{-\frac{1}{2}u^2}\sigma du = e^{4\mu}\frac{1}{\sqrt{2\pi}}$$

$$\int_{-\infty}^{\infty} e^{4u\sigma}e^{-\frac{1}{2}u^2}du = e^{4\mu}\frac{1}{\sqrt{2\pi}}\int_{-\infty}^{\infty}e^{-\frac{1}{2}\left(u^2-8u\sigma\right)}du$$

$$= e^{4\mu}\frac{1}{\sqrt{2\pi}}\int_{-\infty}^{\infty}e^{-\frac{1}{2}(u-4\sigma)^2+\frac{16}{2}\sigma^2}du = e^{4\mu+8\sigma^2}\frac{1}{\sqrt{2\pi}}\int_{-\infty}^{\infty}e^{-\frac{1}{2}(u-4\sigma)^2}du$$

$$Let\ \ p = u - 4\sigma \Rightarrow dp = du$$

Substituting the above results, we get:

$$= e^{4\mu+8\sigma^2}\int_{-\infty}^{\infty}\frac{1}{\sqrt{2\pi}}e^{-\frac{1}{2}p^2}dp$$

As we know that $\int_{-\infty}^{\infty}\frac{1}{\sqrt{2\pi}}e^{-\frac{1}{2}p^2}dp = 1,$ therefore

$$= e^{4\mu+8\sigma^2} \tag{4.6.4}$$

r^th^ Moments of log-normal distribution:

$$\mu_r' = E(X^r) = \int_0^{\infty} x^r f(x)dx = \int_0^{\infty} x^r \frac{1}{x\sqrt{2\pi}\sigma}e^{-\frac{1}{2}\left(\frac{\log_e x-\mu}{\sigma}\right)^2}dx$$

$$Let\ \ u = \frac{\log_e x - \mu}{\sigma} \Rightarrow x = e^{(u\sigma+\mu)}$$

$$\Rightarrow du = \frac{1}{x\sigma} dx \Rightarrow \quad dx = x\sigma du \quad \Rightarrow dx = e^{(u\sigma+\mu)} \sigma du$$

Using the above results, we get:

$$= \int_{-\infty}^{\infty} e^{r(u\sigma+\mu)} \frac{1}{\sqrt{2\pi}\sigma} \mathrm{e}^{-\frac{1}{2}u^2} \sigma du = e^{r\mu} \frac{1}{\sqrt{2\pi}}$$

$$\int_{-\infty}^{\infty} e^{ru\sigma} \mathrm{e}^{-\frac{1}{2}u^2} du = e^{r\mu} \frac{1}{\sqrt{2\pi}} \int_{-\infty}^{\infty} \mathrm{e}^{-\frac{1}{2}\left(u^2-2ru\sigma\right)} du$$

$$= e^{r\mu} \frac{1}{\sqrt{2\pi}} \int_{-\infty}^{\infty} \mathrm{e}^{-\frac{1}{2}(u-r\sigma)^2 + \frac{r^2}{2}\sigma^2} du = e^{r\mu+\frac{r^2}{2}\sigma^2} \frac{1}{\sqrt{2\pi}} \int_{-\infty}^{\infty} \mathrm{e}^{-\frac{1}{2}(u-r\sigma)^2} du$$

$$Let\ \ p = u - r\sigma \Rightarrow dp = du$$

Substituting the above results, we get:

$$= e^{r\mu+\frac{r^2}{2}\sigma^2} \int_{-\infty}^{\infty} \frac{1}{\sqrt{2\pi}} \mathrm{e}^{-\frac{1}{2}p^2} dp$$

As we know that $\int_{-\infty}^{\infty} \frac{1}{\sqrt{2\pi}} \mathrm{e}^{-\frac{1}{2}p^2} dp = 1$, therefore:

$$= e^{r\mu+\frac{r^2}{2}\sigma^2}$$

Substituting the values of r (= 1, 2, 3, 4), we get all four moments which are mentioned in Eqns. (4.6.1)–(4.6.4)

e. **Central Moments of Log Normal Distribution:** The first four central moments $\mu'_1, \mu'_2, \mu'_3, \mu'_4$ have been derived as below:
As we know that the first central is moment is zero.

$$\mu_1 = 0 \tag{4.6.5}$$

$$\mu_2 = \mu'_2 - \mu'^2_1$$

Substituting the values from the Eqns. (4.6.1) and (4.6.2) in the above expression, we get:

$$= e^{2\mu+2\sigma^2} - \left(e^{\mu+\frac{1}{2}\sigma^2}\right)^2$$

$$= e^{2\mu+\sigma^2}\left(e^{\sigma^2}-1\right) \tag{4.6.6}$$

Thus, the *variance* of the uniform discrete distribution is $e^{2\mu}\left(e^{\sigma^2}-1\right)$

$$\mu_3 = \mu_3' - 3\mu_2'\mu_1' + 2\mu_1'^3$$

Substituting the values from the Eqns. (4.6.1) and (4.6.3) in the above expression, we get:

$$= e^{3\mu+\frac{9}{2}\sigma^2} - 3e^{2\mu+2\sigma^2}\, e^{\mu+\frac{1}{2}\sigma^2} + 2\left(e^{\mu+\frac{1}{2}\sigma^2}\right)^3$$

$$= e^{3\mu+\frac{9}{2}\sigma^2} - 3e^{3\mu+\frac{5}{2}\sigma^2} + 2e^{3\mu+\frac{3}{2}\sigma^2}$$

$$= e^{3\mu+\frac{3}{2}\sigma^2}\left(e^{3\sigma^2} - 3e^{\sigma^2} + 2\right)$$

$$= e^{3\mu+\frac{3}{2}\sigma^2}\left(e^{\sigma^2}-1\right)^2\left(e^{\sigma^2}+2\right) \tag{4.6.7}$$

$$\mu_4 = \mu_4' - 4\mu_3'\mu_1' + 6\mu_2'\mu_1'^2 - 3\mu_1'^4,$$

Substituting the values from the Eqns. (4.6.1) and (4.6.4) in the above expression, we get:

$$= e^{4\mu+8\sigma^2} - 4\left(e^{3\mu+\frac{9}{2}\sigma^2}\right)e^{\mu+\frac{1}{2}\sigma^2} + 6\left(e^{2\mu+2\sigma^2}\right)\left(e^{\mu+\frac{1}{2}\sigma^2}\right)^2 - 3\left(e^{\mu+\frac{1}{2}\sigma^2}\right)^4$$

$$= e^{4\mu+8\sigma^2} - 4e^{4\mu+5\sigma^2} + 6e^{4\mu+3\sigma^2} - 3e^{4\mu+2\sigma^2}$$

$$= e^{4\mu+2\sigma^2}\left(e^{6\sigma^2} - 4e^{3\sigma^2} + 6e^{\sigma^2} - 3\right)$$

$$= e^{4\mu+2\sigma^2}\left(e^{\sigma^2}-1\right)^2\left(e^{4\sigma^2} + 2e^{3\sigma^2} + 3e^{2\sigma^2} - 3\right) \tag{4.6.8}$$

f. **Skewness and Kurtosis of Log Normal Distribution:** To find the shape of the distribution, we have derived the expressions of skewness (β_1, γ_1) and kurtosis (β_2, γ_2), which are given below:

$$\beta_1 = \frac{\mu_3^2}{\mu_2^3}$$

Substituting the values from the Eqns. (4.6.6) and (4.6.7) in the above expression, we get:

$$= \frac{\left[e^{3\mu + \frac{3}{2}\sigma^2} \left(e^{\sigma^2} - 1\right)^2 \left(e^{\sigma^2} + 2\right) \right]^2}{\left(e^{2\mu + \sigma^2} \left(e^{\sigma^2} - 1\right) \right)^3}$$

$$= \left(e^{\sigma^2} - 1\right)\left(e^{\sigma^2} + 2\right)^2$$

Using the value of β_1, we get:

$$\gamma_1 = \sqrt{\beta_1}$$

$$\gamma_1 = \left(e^{\sigma^2} + 2\right)\sqrt{\left(e^{\sigma^2} - 1\right)}$$

$$\beta_2 = \frac{\mu_4}{\mu_2^2}$$

Substituting the values from the Eqns. (4.6.6) and (4.6.8) in the above expression, we get:

$$= \frac{e^{4\mu + 2\sigma^2} \left(e^{\sigma^2} - 1\right)^2 \left(e^{4\sigma^2} + 2e^{3\sigma^2} + 3e^{2\sigma^2} - 3\right)}{e^{4\mu + 2\sigma^2} \left(e^{\sigma^2} - 1\right)^2}$$

$$= e^{4\sigma^2} + 2e^{3\sigma^2} + 3e^{2\sigma^2} - 3$$

Using the value of β_2, we get:

$$\gamma_2 = \beta_2 - 3$$

$$\gamma_2 = e^{4\sigma^2} + 2e^{3\sigma^2} + 3e^{2\sigma^2} - 6$$

Note: The other properties of the Log-Normal distribution can be derived utilizing the methods used above for the earlier derivations.

KEYWORDS

- **characteristic function**
- **cumulants generating function**
- **moment generating function**
- **normal distribution**
- **probability generating function**

REFERENCES

Casella, G., & Berger, R. L., (2002). *Statistical Inference*. Duxbury: Belmont, CA.

Gupta, S. C., & Kapoor, V. K., (1997). *Fundamentals of Mathematical Statistics*. Sultan Chand and Sons: New Delhi.

Hogg, R. V., McKean, J. W., & Craig, A. T., (2013). *Introduction to Mathematical Statistics* (7th edn.). Pearson: Boston.

Miller, I., Miller, M., Freund, J. E., & Miller, I., (2004). *John E. Freund's Mathematical Statistics with Applications*. Prentice Hall: Upper Saddle River, NJ.

Rohatgi, V., & Saleh, A. K. Md. E., (2015). *Introduction to Probability and Statistics* (3rd edn.). John Wiley.

CHAPTER 5

Family of Weibull Distributions

MOHD. ARSHAD[1], VIJAY KUMAR[2], MUKTI KHETAN[3], and FOZIA HOMA[4]

[1]*Department of Mathematics, Indian Institute of Technology Indore, Simrol, Indore, Madhya Pradesh, India*

[2]*Department of Mathematics, Indian Institute of Technology Bombay, Powai, Mumbai, Maharashtra, India*

[3]*Department of Mathematics, Amity School of Applied Sciences, Amity University Mumbai, Maharashtra*

[4]*Department of Statistics, Mathematics, and Computer Application, Bihar Agricultural University, Sabour, Bhagalpur, Bihar, India*

5.1 INTRODUCTION

Weibull Distribution was introduced by Waloddi Weibull in the year 1951 and therefore is named after him, although it was first identified by Frechet (1927). Rosin and Rammler (1933) provided an application of this model on the particle size distribution. Weibull model is the most popular lifetime distribution in reliability engineering, and biological studies due to its flexibility, closed-form survival function, and monotonic failure (or hazard) rates. But this model does not provide a good fit to data sets with bathtub-shaped or upside-down bathtub-shaped failure rates. It is evident from the citations (more than 11,900) of the Weibull (1951) that it is a more applicable model to analyze several lifetime

Distribution Theory: Principles and Applications. Fozia Homa, Mukti Khetan, Mohd. Arshad, & Pradeep Mishra (Naz) (Eds.)

events. Modifications of the Weibull model have also been made to extend its applicability to a real-life scenario. In this chapter, we will discuss the Weibull model as well as some modified versions. The Weibull model is also related to several other probability models; e.g., the exponential and Rayleigh models are particular cases of this model. For more details and applications, readers can see the handbook written by Rinne (2009).

➢ **Definition:** A positive random variable X is said to follow Weibull Distribution if its cumulative distribution function (CDF) is

$$F_X(x) = 1 - \exp\left[-\left(\frac{x-a}{b}\right)^c\right]; \qquad x > a \tag{5.1.1}$$

Symbolically: $X \sim W(a, b, c)$

Here a, b, and c are the three parameters of this model in which a is the shift (or location) parameter, b (> 0)is the scale parameter, and c (> 0) is the shape parameter.

The probability density function (PDF) is

$$f_X(x) = \frac{d}{dx}F_X(x) = \frac{d}{dx}\left\{1 - \exp\left[-\left(\frac{x-a}{b}\right)^c\right]\right\}$$

$$f_X(x) = c\frac{1}{b}\left(\frac{x-a}{b}\right)^{c-1} \exp\left[-\left(\frac{x-a}{b}\right)^c\right];$$

$$x > a,\ b > 0,\ c > 0,\ -\infty < a < \infty \tag{5.1.2}$$

Sample Generation of Weibull Distribution

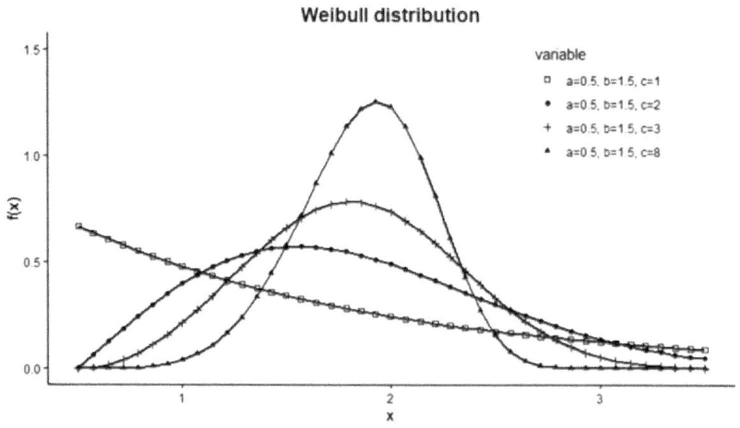

FIGURE 5.1 Weibull distribution.

R Code: Generate random sample from Weibull distribution.

x = rweibull(n, c, b)

where; n is the sample size; b, c are the parameters of the distribution

Example:

x = rweibull(10, 2, 2)

```
Console   Terminal   Jobs

> x= rweibull(10, 2, 2)
> x
 [1] 2.775109 1.565645 1.569429 0.798047 1.999497 1.049338 4.378570 1.249547 1.380271 1.581562
>
```

If $a \neq 0$

- **Step 1:** Generate x= rweibull(n, c, b)
- **Step 2:** Generate $y = x + a$

i.e., y denotes the vector containing W(a, b, c) data set of size n.

Sample Generation of Using Inverse Transformation Method:

$$F_X(x) = 1 - \exp\left[-\left(\frac{X-a}{b}\right)^c\right]; \qquad x > a$$

$$U = 1 - \exp\left[-\left(\frac{X-a}{b}\right)^c\right]$$

$$\log(1-U) = -\left(\frac{X-a}{b}\right)^c$$

$$X = b\left[-\log(1-U)\right]^{1/c} + a$$

where; U follows uniform (0,1).

Note: First generate n data points $\underline{u} = (u_1, u_2, \ldots, u_n)$ from $U(0,1)$ distribution. Now, consider the transformation $x_i = b[-\log(1-u_i)]^{1/c} + a$; $i = 1, 2, \ldots n$. Then, the data set $\underline{x} = (x_1, x_2, \ldots, x_n)$ has $W(a, b, c)$ distribution.

- **Example 5.1:** Let u = (0.1,0.3,0.22,0.51,0.11,0.6) be a data set generated from $U(0,1)$ distribution. Now, we want to generate random samples from W (1, 2, 2), W (2, 1, 3) and W (1, 1, 2). Calculate $x_i = b[-\log(1-u_i)]^{1/c} + a$; $i = 1,2,\ldots 6$, where $u_i = (u_1, u_2, \ldots, u_6)$.

U(0,1)	Random Sample		
	W(1, 1, 2)	W(1, 2, 2)	W(2, 1, 3)
0.10	1.324593	1.649186	2.472309
0.30	1.597223	2.194445	2.709182
0.22	1.498459	1.996918	2.628665
0.51	1.844600	2.689201	2.893513
0.11	1.341370	1.682741	2.488447
0.60	1.957231	2.914462	2.971280

5.1.1 PROPERTIES OF WEIBULL DISTRIBUTION

a. **Raw Moments of Weibull Distribution:** In this section, we derived the moments about zero known as *raw moments*. The first four raw moments $\mu_1', \mu_2', \mu_3', \mu_4'$ have been derived as below:

$$\mu_1' = E(X) = \int_{-\infty}^{\infty} xf(x)dx = c\frac{1}{b}\int_a^{\infty} x\left(\frac{x-a}{b}\right)^{c-1} \exp\left[-\left(\frac{x-a}{b}\right)^c\right] dx$$

$$\text{Let } \left(\frac{x-a}{b}\right)^c = z \Rightarrow x = bz^{1/c} + a \Rightarrow dx = b\frac{1}{c}z^{(1/c)-1}dz$$

$$\mu_1' = c\frac{1}{b}\int_0^{\infty}\left(bz^{1/c} + a\right)z^{(1-(1/c))} \exp(-z) b\frac{1}{c}z^{(1/c)-1}dz$$

$$= \int_0^{\infty}\left(bz^{1/c} + a\right)\exp(-z)dz$$

$$= b\int_0^{\infty} z^{1/c} \exp(-z)dz + a\int_0^{\infty} \exp(-z)dz$$

Using the usual gamma function, we get

$$\mu_1' = b\overline{\left|(1+(1/c))\right.} + a. \tag{5.1.3}$$

So, the *mean* of the Weibull distribution is $\mu_1' = b\overline{\left|(1+(1/c))\right.} + a$ and if $(1/c)$ it is an integer, the mean is $b(1/c)! + a$.

The second raw moment is:

$$\mu_2' = E\left(X^2\right) = \int_{-\infty}^{\infty} x^2 f(x)dx = c\frac{1}{b}\int_a^{\infty} x^2\left(\frac{x-a}{b}\right)^{c-1} \exp\left[-\left(\frac{x-a}{b}\right)^c\right] dx$$

$$\text{Let } \left(\frac{x-a}{b}\right)^c = z \Rightarrow x = bz^{1/c} + a \Rightarrow dx = b\frac{1}{c}z^{(1/c)-1}dz$$

$$\mu_2' = c\frac{1}{b}\int_0^\infty \left(bz^{1/c}+a\right)^2 z^{(1-(1/c))}\exp(-z)b\frac{1}{c}z^{(1/c)-1}dz$$

$$= \int_0^\infty \left(bz^{1/c}+a\right)^2 \exp(-z)dz$$

$$= \int_0^\infty \left(b^2z^{2/c}+a^2+2abz^{1/c}\right)\exp(-z)dz$$

$$= \int_0^\infty b^2z^{2/c}\exp(-z)dz + \int_0^\infty a^2\exp(-z)dz + 2ab\int_0^\infty z^{1/c}\exp(-z)dz$$

Using the result of the gamma function, we get

$$\mu_2' = b^2\overline{\left|1+(2/c)\right.} + a^2 + 2ab\overline{\left|1+(1/c)\right.} \tag{5.1.4}$$

The third raw moment is

$$\mu_3' = E\left(X^3\right) = \int_{-\infty}^{\infty} x^3 f(x)dx = c\frac{1}{b}\int_a^\infty x^3\left(\frac{x-a}{b}\right)^{c-1}\exp\left[-\left(\frac{x-a}{b}\right)^c\right]dx$$

$$\text{Let } \left(\frac{x-a}{b}\right)^c = z \Rightarrow x = bz^{1/c} + a \Rightarrow dx = b\frac{1}{c}z^{(1/c)-1}dz$$

$$\mu_3' = c\frac{1}{b}\int_0^\infty \left(bz^{1/c}+a\right)^3 z^{(1-(1/c))}\exp(-z)b\frac{1}{c}z^{(1/c)-1}dz$$

$$= \int_0^\infty \left(bz^{1/c}+a\right)^3 \exp(-z)dz$$

$$= \int_0^\infty \left(b^3z^{3/c}+a^3+3a^2bz^{1/c}+3ab^2z^{2/c}\right)\exp(-z)dz$$

$$= b^3\int_0^\infty z^{3/c}\exp(-z)dz + \int_0^\infty a^3\exp(-z)dz + 3a^2$$

$$b\int_0^\infty z^{1/c}\exp(-z)dz + 3ab^2\int_0^\infty z^{2/c}\exp(-z)dz$$

Using the result of the gamma function, we get

$$\mu_3' = b^3\overline{\left|1+(3/c)\right.} + a^3 + 3a^2b\overline{\left|1+(1/c)\right.} + 3ab^2\overline{\left|1+(2/c)\right.} \tag{5.1.5}$$

The fourth raw moment is

$$\mu_4' = E\left(X^4\right) = \int_{-\infty}^{\infty} x^4 f(x)dx = c\frac{1}{b}\int_a^\infty x^4\left(\frac{x-a}{b}\right)^{c-1}\exp\left[-\left(\frac{x-a}{b}\right)^c\right]dx$$

$$\text{Let } \left(\frac{x-a}{b}\right)^c = z \Rightarrow x = bz^{1/c} + a \Rightarrow dx = b\frac{1}{c}z^{(1/c)-1}dz$$

$$\mu_4' = c\frac{1}{b}\int_0^\infty \left(bz^{1/c}+a\right)^4 z^{(1-(1/c))}\exp(-z)b\frac{1}{c}z^{(1/c)-1}dz$$

$$= \int_0^\infty \left(bz^{1/c}+a\right)^4 \exp(-z)dz$$

$$= \int_0^\infty \left(b^4z^{4/c} + a^4 + 4a^3bz^{1/c} + 4ab^3z^{3/c} + 6a^2b^2z^{2/c}\right)\exp(-z)dz$$

$$= \left[\begin{array}{l} b^4\int_0^\infty z^{4/c}\exp(-z)dz + \int_0^\infty a^4\exp(-z)dz + 4a^3b\int_0^\infty z^{1/c}\exp(-z)dz \\ +4ab^3\int_0^\infty z^{3/c}\exp(-z)dz + 6a^2b^2\int_0^\infty z^{2/c}\exp(-z)dz \end{array}\right]$$

Using the result of the gamma function, we get

$$\begin{aligned}\mu_4' &= b^4\Gamma\left(1+(4/c)\right) + a^4 + 4a^3b\Gamma\left(1+(1/c)\right) \\ &\quad + 4ab^3\Gamma\left(1+(3/c)\right) + 6a^2b^2\Gamma\left(1+(2/c)\right)\end{aligned} \qquad (5.1.6)$$

b. **Central Moments of Weibull Distribution:** Central moments are very essential for the distribution as they tell the characteristics of the distribution. So, the first four central moments μ_1, μ_2, μ_3, μ_4 have been derived as below.
We know that the first central moment is always zero, i.e.,

$$\mu_1 = 0 \qquad (5.1.7)$$

The second central moment tells the variation of the distribution and is known as the **variance**. The second central moment is given below.

$$\mu_2 = \mu_2' - \mu_1'^2$$

Using the values from Eqns. (5.1.3) and (5.1.4), we get:

$$\mu_2 = b^2\Gamma\left(1+(2/c)\right) + a^2 + 2ab\Gamma\left(1+(1/c)\right) - \left(b\Gamma\left(1+(1/c)\right) + a\right)^2$$

$$= b^2\Gamma\left(1+(2/c)\right) + a^2 + 2ab\Gamma\left(1+(1/c)\right) - b^2\left\{\Gamma\left(1+(1/c)\right)\right\}^2 - a^2 - 2ab\Gamma\left(1+(1/c)\right)$$

$$\mu_2 = b^2\left[\Gamma\left(1+(2/c)\right) - \left\{\Gamma\left(1+(1/c)\right)\right\}^2\right] \tag{5.1.8}$$

Thus, the *variance* of the *Weibull distribution* is $b^2\left[\Gamma\left(1+(2/c)\right) - \left\{\Gamma\left(1+(1/c)\right)\right\}^2\right]$.

The third central moment tells the skewness of the distribution. The derivation of the third central moment is given below:

$$\mu_3 = \mu_3' - 3\mu_2'\mu_1' + 2\mu_1'^3$$

Using the values from Eqns. (5.1.3) to (5.1.5), we get:

$$= \begin{bmatrix} b^3\Gamma\left(1+(3/c)\right) + a^3 + 3a^2b\Gamma\left(1+(1/c)\right) + 3ab^2\Gamma\left(1+(2/c)\right) \\ -3\left[\left(b^2\Gamma\left(1+(2/c)\right) + a^2 + 2ab\Gamma\left(1+(1/c)\right)\right)\left(b\Gamma\left(1+(1/c)\right) + a\right)\right] + 2\left[b\Gamma\left(1+(1/c)\right) + a\right]^3 \end{bmatrix}$$

$$= \begin{bmatrix} b^3\Gamma\left(1+(3/c)\right) + a^3 + 3a^2b\Gamma\left(1+(1/c)\right) + 3ab^2\Gamma\left(1+(2/c)\right) \\ -3\left\{\begin{matrix} b^3\Gamma\left(1+(2/c)\right)\Gamma\left(1+(1/c)\right) + a^2b\Gamma\left(1+(1/c)\right) + 2ab^2\left\{\Gamma\left(1+(1/c)\right)\right\}^2 \\ +ab^2\Gamma\left(1+(2/c)\right) + a^3 + 2a^2b\Gamma\left(1+(1/c)\right) \end{matrix}\right\} \\ +2\left\{b^3\left\{\Gamma\left(1+(1/c)\right)\right\}^3 + a^3 + 3a^2b\Gamma\left(1+(1/c)\right) + 3ab^2\left\{\Gamma\left(1+(1/c)\right)\right\}^2\right\} \end{bmatrix}$$

$$\mu_3 = b^3\left[\Gamma\left(1+(3/c)\right) - 3\Gamma\left(1+(2/c)\right)\Gamma\left(1+(1/c)\right) + 2\left\{\Gamma\left(1+(1/c)\right)\right\}^3\right] \tag{5.1.9}$$

The fourth central moment helps to find the kurtosis of the distribution. The fourth central moment is given below:

$$\mu_4 = \mu_4' - 4\mu_3'\mu_1' + 6\mu_2'\mu_1'^2 - 3\mu_1'^4$$

Using the values from Eqns. (5.1.3) to (5.1.6), we get:

$$= \begin{bmatrix} b^4\Gamma\left(1+(4/c)\right) + a^4 + 4a^3b\Gamma\left(1+(1/c)\right) + 4ab^3\Gamma\left(1+(3/c)\right) + 6a^2b^2\Gamma\left(1+(2/c)\right) \\ -4\left(b^3\Gamma\left(1+(3/c)\right) + a^3 + 3a^2b\Gamma\left(1+(1/c)\right) + 3ab^2\Gamma\left(1+(2/c)\right)\right)\left(b\Gamma\left(1+(1/c)\right) + a\right) \\ +6\left\{\left(b^2\Gamma\left(1+(2/c)\right) + a^2 + 2ab\Gamma\left(1+(1/c)\right)\right)\left(b\Gamma\left(1+(1/c)\right) + a\right)^2\right\} - 3\left\{b\Gamma\left(1+(1/c)\right) + a\right\}^4 \end{bmatrix}$$

$$
= \left[\begin{array}{l}
b^4\Gamma(1+(4/c)) + a^4 + 4a^3b\Gamma(1+(1/c)) + 4ab^3\Gamma(1+(3/c)) + 6a^2b^2\Gamma(1+(2/c)) \\
-4\left(\begin{array}{l} b^4\Gamma(1+(3/c))\Gamma(1+(1/c)) + a^3b\Gamma(1+(1/c)) + 3a^2b^2\left\{\Gamma(1+(1/c))\right\}^2 + 3ab^3\Gamma(1+(2/c))\Gamma(1+(1/c)) \\ +ab^3\Gamma(1+(3/c)) + a^4 + 3a^3b\Gamma(1+(1/c)) + 3a^2b^2\Gamma(1+(2/c)) \end{array} \right) \\
+6\left(\begin{array}{l} b^4\Gamma(1+(2/c))\left\{\Gamma(1+(1/c))\right\}^2 + a^2b^2\left\{\Gamma(1+(1/c))\right\}^2 + 2ab^3\left\{\Gamma(1+(1/c))\right\}^3 \\ +a^2b^2\Gamma(1+(2/c)) + a^4 + 2a^3b\Gamma(1+(1/c)) \\ +2ab^3\Gamma(1+(2/c))\Gamma(1+(1/c)) + 2a^3b\Gamma(1+(1/c)) + 4a^2b^2\left\{\Gamma(1+(1/c))\right\}^2 \end{array} \right) \\
-3\left(b^4\left\{\Gamma(1+(1/c))\right\}^4 + a^4 + 4a^3b\Gamma(1+(1/c)) + 4ab^3\left\{\Gamma(1+(1/c))\right\}^3 + 6a^2b^2\left\{\Gamma(1+(1/c))\right\}^2 \right)
\end{array} \right]
$$

$$
\mu_4 = b^4 \left(\begin{array}{l} \Gamma(1+(4/c)) - 4\Gamma(1+(3/c))\Gamma(1+(1/c)) + 6\Gamma(1+(2/c)) \\ \left\{\Gamma(1+(1/c))\right\}^2 - 3\left\{\Gamma(1+(1/c))\right\}^4 \end{array} \right) \tag{5.1.10}
$$

c. **Skewness and Kurtosis of Weibull Distribution:** To find the shape of this model, we have derived the expressions of skewness (β_1, γ_1) and kurtosis (β_2, γ_2) in this section.

$$
\beta_1 = \frac{\mu_3^2}{\mu_2^3}
$$

Using the values from Eqns. (5.1.8) and (5.1.9), we get:

$$
\beta_1 = \frac{\left[\Gamma(1+(3/c)) - 3\Gamma(1+(2/c))\Gamma(1+(1/c)) + 2\left\{\Gamma(1+(1/c))\right\}^3\right]^2}{\left[\Gamma(1+(2/c)) - \left\{\Gamma(1+(1/c))\right\}^2\right]^3} \tag{5.1.11}
$$

$$
\gamma_1 = \sqrt{\beta_1} = \frac{\mu_3}{\mu_2^{3/2}}
$$

Substituting the value of β_1 in the above expression, we get:

$$
\gamma_1 = \frac{\Gamma(1+(3/c)) - 3\Gamma(1+(2/c))\Gamma(1+(1/c)) + 2\left\{\Gamma(1+(1/c))\right\}^3}{\left[\Gamma(1+(2/c)) - \left\{\Gamma(1+(1/c))\right\}^2\right]^{3/2}} \tag{5.1.12}
$$

$$
\beta_2 = \frac{\mu_4}{\mu_2^2}
$$

Using the values from Eqns. (5.1.8) and (5.1.10), we get:

$$\beta_2 = \frac{\Gamma\left(1+(4/c)\right) - 4\Gamma\left(1+(3/c)\right)\Gamma\left(1+(1/c)\right) + 6\Gamma\left(1+(2/c)\right)\left\{\Gamma\left(1+(1/c)\right)\right\}^2 - 3\left\{\Gamma\left(1+(1/c)\right)\right\}^4}{\left[\Gamma\left(1+(2/c)\right) - \left\{\Gamma\left(1+(1/c)\right)\right\}^2\right]^2} \qquad (5.1.13)$$

Substituting the value of β_2 in the above expression, we get:

$$\gamma_2 = \frac{\Gamma\left(1+(4/c)\right) - 4\Gamma\left(1+(3/c)\right)\Gamma\left(1+(1/c)\right) + 6\Gamma\left(1+(2/c)\right)\left\{\Gamma\left(1+(1/c)\right)\right\}^2 - 3\left\{\Gamma\left(1+(1/c)\right)\right\}^4}{\left[\Gamma\left(1+(2/c)\right) - \left\{\Gamma\left(1+(1/c)\right)\right\}^2\right]^2} - 3 \qquad (5.1.14)$$

➢ **Shapes of the Weibull Distribution at Different Values of Parameters**

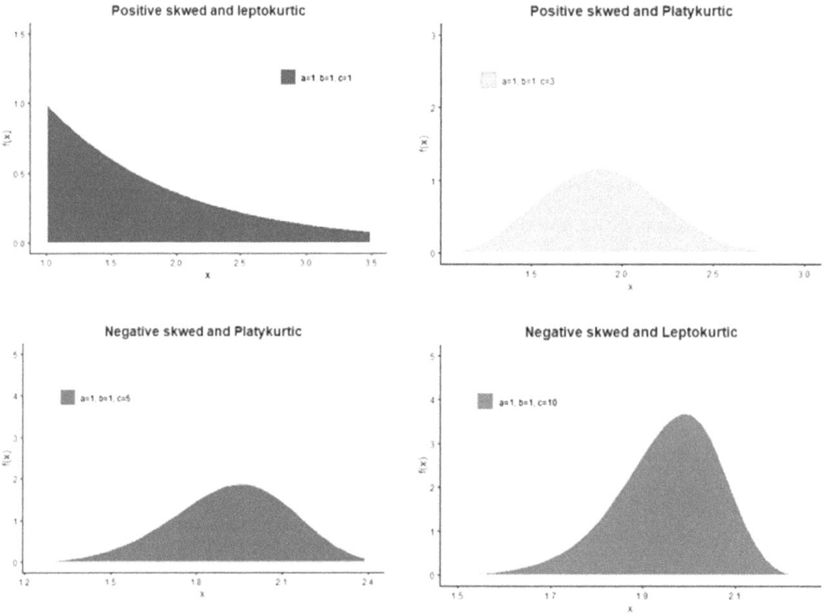

FIGURE 5.2 Shape of the Weibull distribution.

d. Median, Quartile, and Percentile of Weibull Distribution: Median divides the data into two equal parts or in other words we can write

$$F_X(M) = \frac{1}{2},$$

where M denotes the median.

From the above section we know the CDF of Weibull distribution, we get

$$1-\exp\left[-\left(\frac{M-a}{b}\right)^{c}\right]=\frac{1}{2}$$

$$\exp\left[-\left(\frac{M-a}{b}\right)^{c}\right]=\frac{1}{2}\Rightarrow-\left(\frac{M-a}{b}\right)^{c}=-\log_e 2$$

$$\Rightarrow M=b\left(\log_e 2\right)^{1/c}+a \qquad (5.1.15)$$

Quartiles divide the data into four equal parts or in other words we can write

$$F_X(Q_r)=\frac{r}{4}, \qquad r=1,2,3$$

where Q_r denotes the rth quantile.

$$Q_r=b\left(-\log_e\frac{r}{4}\right)^{1/c}+a$$

Clearly, Q_2 equals to M, i.e., the second quantile is equal to the median.

Percentiles divide the data into 100 equal parts or in other words we can write

$$F_X(P_r)=\frac{r}{100} \qquad r=1,2,\ldots,99$$

where P_r denotes the rth percentile.

$$P_r=b\left(-\log_e\frac{r}{100}\right)^{1/c}+a \qquad r=1,2,\ldots,99$$

We can find the expressions of P_1, P_2, …, P_{100} by putting r =1 to 99, respectively in the above expression.

e. Mode of Weibull Distribution: The behavior of the PDF of the Weibull model depends on parameter c. For obtaining the mode of this model, we will consider the following two cases.

➢ **Case 1: when c is not equal to 1.** Mode is the highest occurred value in the data set and to find the mode of the distribution we have to differentiate the PDF with respect to x which is given below:

$$\log f_X(x)=\frac{d}{dx}\log_e\left(c\frac{1}{b}\left(\frac{x-a}{b}\right)^{c-1}\exp\left[-\left(\frac{x-a}{b}\right)^{c}\right]\right)$$

$$\log f_X(x) = \frac{d}{dx}\left(\log_e c - \log_e b + (c-1)\log_e\left(\frac{x-a}{b}\right) - \left(\frac{x-a}{b}\right)^c\right)$$

Differentiating both the side with respect to x

$$\frac{1}{f_X(x)} f_X'(x) = 0 + 0 + (c-1)\frac{1}{\left(\frac{x-a}{b}\right)}\frac{1}{b} - c\left(\frac{x-a}{b}\right)^{c-1}\frac{1}{b}$$

$$f_x'(x) = \left[(c-1)\frac{1}{(x-a)} - c\frac{1}{b}\left(\frac{x-a}{b}\right)^{c-1}\right] f_X(x)$$

Clearly, $f_x'(x) = 0 \; if \left[(c-1)\frac{1}{(x-a)} - c\frac{1}{b}\left(\frac{x-a}{b}\right)^{c-1}\right] f_X(x) = 0$.

Since PDF is always positive, we get:

$$\Rightarrow (c-1)\frac{1}{(x-a)} - c\frac{1}{b}\left(\frac{x-a}{b}\right)^{c-1} = 0; \; c \neq 1$$

$$(c-1)\frac{1}{(x-a)} = c\frac{1}{b}\left(\frac{x-a}{b}\right)^{c-1} \Rightarrow \frac{c-1}{c} = \left(\frac{x-a}{b}\right)^c$$

$$x = b\left(\frac{c-1}{c}\right)^{1/c} + a; \qquad c \neq 1 \tag{5.1.16}$$

Thus, for $c \neq 1$, the mode of the distribution is given by Eqn. (5.1.16).

Case 2: When $c = 1$. In this case, the PDF of this model is

$$f_X(x) = \frac{1}{b}\exp\left[-\left(\frac{x-a}{b}\right)\right]; x > a \tag{5.1.17}$$

The PDF $f(x)$, given in (5.1.17), is a decreasing function of x. Thus, the mode of this model occurs at the beginning of the support of x, i.e., the mode the distribution is $x = a$.

f. **Reliability Function of Weibull Distribution:** In reliability analysis, the random variable of interest is the failure time of the component/system. Let us denote this random variable by X. Then, by definition, reliability $R(x)$ at the time x, of the component/system is given by

   $R(x)$ = Probability that the time to failure of the component/system is greater than x

$$R(x) = P[X > x] = 1 - F_x(x)$$

Using Eqn. (5.1.1), we get:

$$R(x) = \begin{cases} \exp\left[-\left(\frac{x-a}{b}\right)^c\right] & ;x > a \\ 1 & ;x \le a \end{cases} \tag{5.1.18}$$

g. Hazard Rate of Weibull Distribution: The instantaneous rate of failure at time x is known as the hazard rate and is defined by

$$\lambda(x) = \frac{1}{R(x)} f(x).$$

Substituting the expression of R(x) and f(x), we get:

$$\lambda(x) = \frac{1}{\exp\left[-\left(\frac{x-a}{b}\right)^c\right]} \mathrm{c}\frac{1}{b}\left(\frac{x-a}{b}\right)^{c-1} \exp\left[-\left(\frac{x-a}{b}\right)^c\right]$$

$$\lambda(x) = \mathrm{c}\frac{1}{b}\left(\frac{x-a}{b}\right)^{c-1} \quad ;x > a \tag{5.1.19}$$

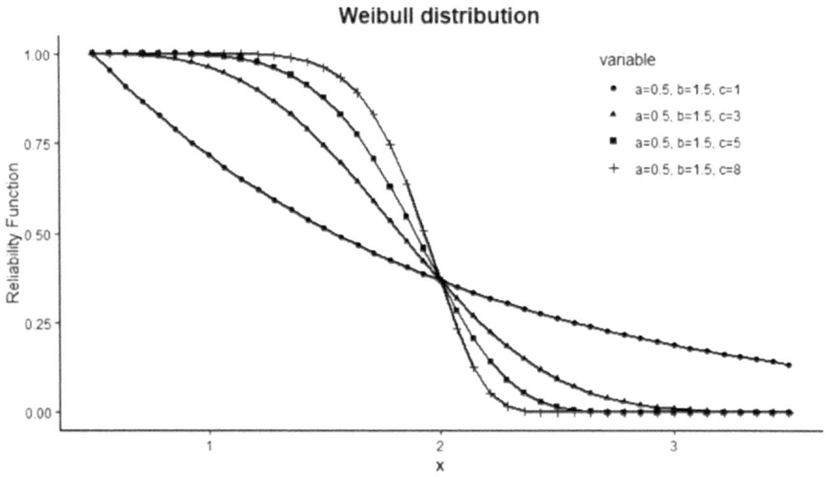

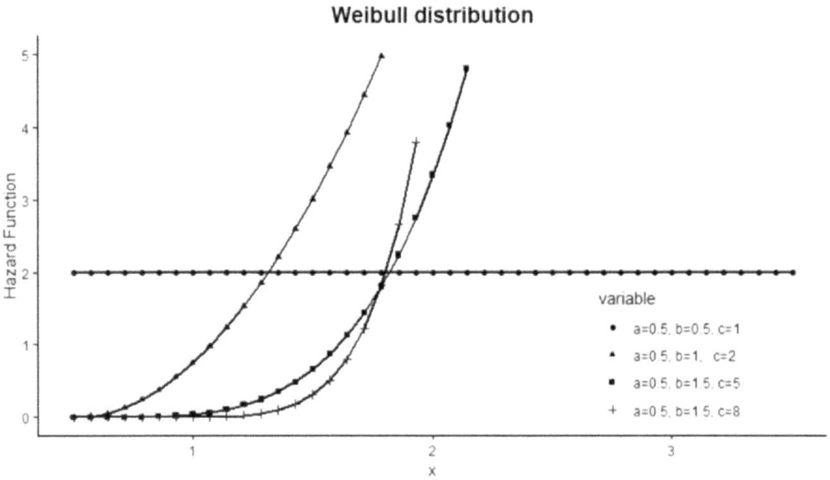

FIGURE 5.3 Reliability function and hazard function of Weibull distribution.

h. Mean Time to Failure of Weibull Distribution: In reliability terminology, the mean of the random variable X in the absence of repair and replacement is known as the meantime to failure (MTTF). The expression of MTTF is given by

$$\text{MTTF} = \int_{-\infty}^{\infty} xf(x)dx = \text{ Mean of the Weibull distribution}$$

Using Eqn. (5.1.3), we get

$$\text{MTTF} = b\Gamma\left(1+(1/c)\right)+a \tag{5.1.20}$$

i. Moment Generating Function (MGF) of Weibull Distribution: The moment generating function (MGF) of this model is given by

$$M_X(t) = E\left(e^{tX}\right) = \int_{-\infty}^{\infty} e^{tx} f(x)dx = c\frac{1}{b}\int_{a}^{\infty} e^{tx}\left(\frac{x-a}{b}\right)^{c-1} \exp\left[-\left(\frac{x-a}{b}\right)^{c}\right] dx$$

$$Let \left(\frac{x-a}{b}\right)^{c} = z \Rightarrow x = bz^{1/c} + a \Rightarrow dx = b\frac{1}{c}z^{(1/c)-1}dz$$

$$M_X(\mathrm{t}) = c\frac{1}{b}\int_{0}^{\infty} e^{t\left(bz^{1/c}+a\right)} z^{(1-(1/c))} \exp(-z) b\frac{1}{c}z^{(1/c)-1}dz$$

$$= e^{ta}\int_0^\infty e^{tbz^{1/c}} \exp(-z)dz$$

$$= e^{ta}\int_0^\infty \sum_{r=0}^{\infty} \frac{\left(tbz^{1/c}\right)^r}{r!} \exp(-z)dz$$

$$= e^{ta}\sum_{r=0}^{\infty} \frac{(tb)^r}{r!} \int_0^\infty z^{r/c} \exp(-z)dz,$$

Using gamma function, we get

$$M_X(t) = e^{ta}\sum_{r=0}^{\infty} \frac{(tb)^r}{r!} \Gamma\left(1+(r/c)\right) \tag{5.1.21}$$

j. Characteristic Function (CF) of Weibull Distribution: The characteristic function (CF) is here. It is denoted by $\phi_X(t)$. The derivation of $\phi_X(t)$ is like the derivation $M_X(t)$, given in Section (i).

$$\phi_X(t) = e^{ita}\sum_{r=0}^{\infty} \frac{(itb)^r}{r!} \Gamma\left(1+(r/c)\right) \tag{5.1.22}$$

k. Entropy of Weibull Distribution: Entropy is a measure of uncertainty. The concept of information entropy was introduced by Shannon in 1948. It is defined as

$$\text{Entropy} = E[-\log f(X)]$$

$$= E\left[-\log_e\left(c\frac{1}{b}\left(\frac{X-a}{b}\right)^{c-1} \exp\left[-\left(\frac{X-a}{b}\right)^c\right]\right)\right]$$

$$= -\log_e\frac{c}{b} - (c-1)E\left[\log_e\left(\frac{X-a}{b}\right)\right] + E\left[\left(\frac{X-a}{b}\right)^c\right] \tag{5.1.23}$$

Now,

$$E\left(\log\left(\frac{X-a}{b}\right)\right) = \int_a^\infty \log\left(\frac{x-a}{b}\right) f(x)dx$$

$$= \int_a^\infty \log\left(\frac{x-a}{b}\right) c\frac{1}{b}\left(\frac{x-a}{b}\right)^{c-1} \exp\left[-\left(\frac{x-a}{b}\right)^c\right]dx$$

$$Let\ \left(\frac{x-a}{b}\right)^{c} = z \Rightarrow\ x = bz^{1/c} + a \Rightarrow dx = b\frac{1}{c}z^{(1/c)-1}dz$$

$$E\left(\log\left(\frac{X-a}{b}\right)\right) = c\frac{1}{b}\int_{0}^{\infty}\log\left(z^{1/c}\right)z^{(1-(1/c))}\exp(-z)b\frac{1}{c}z^{(1/c)-1}dz$$

$$= \int_{0}^{\infty}\log\left(z^{1/c}\right)\exp(-z)dz$$

$$= \frac{1}{c}\int_{0}^{\infty}\log z\exp(-z)dz$$

$$E\left(\log\left(\frac{X-a}{b}\right)\right) = -\frac{\Upsilon}{\mathrm{c}}, \tag{5.1.24}$$

where; $\Upsilon = \int_{0}^{\infty}\log z\exp(-z)dz$ is Euler's Constant.

Consider

$$E\left(\frac{X-a}{b}\right)^{c} = \int_{a}^{\infty}\left(\frac{x-a}{b}\right)^{c} c\frac{1}{b}\left(\frac{x-a}{b}\right)^{c-1}\exp\left[-\left(\frac{x-a}{b}\right)^{c}\right]dx$$

$$Let\ \left(\frac{x-a}{b}\right)^{c} = z \Rightarrow\ x = bz^{1/c} + a \Rightarrow dx = b\frac{1}{c}z^{(1/c)-1}dz$$

$$E\left(\left(\frac{X-a}{b}\right)^{c}\right) = c\frac{1}{b}\int_{0}^{\infty}z^{(2-(1/c))}\exp(-z)b\frac{1}{c}z^{(1/c)-1}dz$$

$$= \int_{0}^{\infty}z\exp(-z)dz$$

$$= \overline{|2}$$

$$E\left(\left(\frac{X-a}{b}\right)^{c}\right) = 1 \tag{5.1.25}$$

Now, using Eqns. (5.1.23) and (5.1.24) in Eqn. (5.1.22), we get:

$$\text{Entropy} = \log_{e}\frac{b}{c} + (c-1)\frac{\Upsilon}{\mathrm{c}} + 1$$

5.2 SPECIAL CASES OF WEIBULL DISTRIBUTION

5.2.1 TWO-PARAMETER WEIBULL DISTRIBUTION

Two-parameter Weibull distribution can be derived by putting a = 0 in eq (5.1.1) and (5.1.2). The CDF and PDF of this model are respectively given by

$$F_X(x) = 1 - \exp\left[-\left(\frac{x}{b}\right)^c\right];\ x > 0,\ b > 0,\ c > 0 \tag{5.2.1}$$

$$f_X(x) = c\frac{1}{b}\left(\frac{x}{b}\right)^{c-1} \exp\left[-\left(\frac{x}{b}\right)^c\right];\ x > 0,\ b > 0,\ c > 0 \tag{5.2.2}$$

Symbolically: $X \sim W(b, c)$

The properties of the two-parameter Weibull distribution are given below.

5.2.1.1 PROPERTIES OF TWO-PARAMETER WEIBULL DISTRIBUTION

a. First Four Moments of this Model are given below:

$$\mu_1' = b\Gamma(1 + (1/c)) \tag{5.2.1.3}$$

$$\mu_2' = b^2\Gamma 1 + (2/c) \tag{5.2.1.4}$$

$$\mu_3' = b^3\Gamma 1 + (3/c) \tag{5.2.1.5}$$

$$\mu_4' = b^4\Gamma 1 + (4/c) \tag{5.2.1.6}$$

b. The first four central moments of two-parameter Weibull model are same as three-parameter Weibull model (given in Eqns. (5.1.17)–(5.1.10))

c. Skewness and kurtosis of two-parameter Weibull model are same as three-parameter Weibull model (given in Eqns. (5.1.11)–(5.1.14))

d. Median, Quartile, Percentile, and Mode of Two-Parameter Weibull Distribution:

$$\text{Median} = b(\log_e 2)^{1/c} \tag{5.2.1.7}$$

$$Q_r = b\left(-\log_e \frac{r}{4}\right)^{1/c} \quad ;r = 1,2,3$$

$$P_r = b\left(-\log_e \frac{r}{100}\right)^{1/c} + a \quad ;r = 1,2,...,99$$

$$\text{Mode} = b\left(\frac{c-1}{c}\right)^{1/c} \tag{5.2.1.8}$$

e. Reliability Function, Hazard Function, and MTTF of Two-Parameter Weibull Distribution

$$R(x) = \exp\left[-\left(\frac{x}{b}\right)^{c}\right];x > 0, b > 0, c > 0 \tag{5.2.1.9}$$

$$\lambda(x) = \mathrm{c}\frac{1}{b}\left(\frac{x}{b}\right)^{c-1} \quad ;x > 0, b > 0, c > 0 \tag{5.2.1.10}$$

$$\text{MTTF} = b\Gamma\left(1+(1/c)\right) \tag{5.2.1.11}$$

f. Moments Generating Function, Characteristic Function (CF) and Entropy of two-parameter Weibull Distribution

$$\mathrm{M_X(t)}=\sum_{r=0}^{\infty}\frac{(tb)^r}{r!}\Gamma\left(1+(\mathrm{r/c})\right) \tag{5.2.1.12}$$

$$\mathrm{M_X(t)}=\sum_{r=0}^{\infty}\frac{(itb)^r}{r!}\Gamma\left(1+(\mathrm{r/c})\right) \tag{5.2.1.13}$$

$$\text{Entropy} = \log_e \frac{b}{c} + (c-1)\frac{\Upsilon}{\mathrm{c}} + 1 \tag{5.2.1.14}$$

5.2.2 ONE PARAMETER WEIBULL DISTRIBUTION

One parameter Weibull distribution can be derived by putting a = 0 and b=1 in eq (5.1.1) and (5.1.2). The CDF and PDF of one parameter Weibull distribution are respectively given by

$$F_X(x) = 1 - \exp\left(-x^c\right);x > 0, c > 0 \tag{5.2.2.1}$$

$$f_X(x) = c(x)^{c-1}\exp\left[-(x)^c\right] \; ; \; x > 0, \; c > 0 \tag{5.2.2.2}$$

Symbolically: $X \sim W(b, c)$

The properties of one parameter Weibull distribution are given below.

5.2.2.1 PROPERTIES OF THE ONE PARAMETER WEIBULL DISTRIBUTION

a. Moments:

$$\mu_1' = \Gamma\left(1+(1/c)\right) \qquad (5.2.2.3)$$

$$\mu_2' = \Gamma\left(1+(2/c)\right) \qquad (5.2.2.4)$$

$$\mu_3' = \Gamma\left(1+(3/c)\right) \qquad (5.2.2.5)$$

$$\mu_4' = \Gamma\left(1+(4/c)\right) \qquad (5.2.2.6)$$

b. Central Moments:

$$\mu_1 = 0 \qquad (5.2.2.7)$$

$$\mu_2 = \Gamma\left(1+(2/c)\right) - \left\{\Gamma\left(1+(1/c)\right)\right\}^2 \qquad (5.2.2.8)$$

$$\mu_3 = \Gamma\left(1+(3/c)\right) - 3\Gamma\left(1+(2/c)\right)\Gamma\left(1+(1/c)\right) + 2\left\{\Gamma\left(1+(1/c)\right)\right\}^3 \qquad (5.2.2.9)$$

$$\mu_4 = \Gamma\left(1+(4/c)\right) - 4\Gamma\left(1+(3/c)\right)\Gamma\left(1+(1/c)\right) + 6\Gamma\left(1+(2/c)\right)\left\{\Gamma\left(1+(1/c)\right)\right\}^2 - 3\left\{\Gamma\left(1+(1/c)\right)\right\}^4 \qquad (5.2.2.10)$$

c. Skewness and kurtosis of one parameter Weibull model are same as three-parameter Weibull model (given in Eqns. (5.1.11)–(5.1.14))

d. Median, Quartile, Percentile, and Mode:

$$\text{Median} = \left(\log_e 2\right)^{1/c} \qquad (5.2.2.11)$$

$$Q_r = \left(-\log_e \frac{r}{4}\right)^{1/c} \quad ;r = 1,2,3$$

$$P_r = \left(-\log_e \frac{r}{100}\right)^{1/c} + a \quad ;r = 1,2,\ldots,99$$

$$\text{Mode} = \left(\frac{c-1}{c}\right)^{1/c} \qquad (5.2.2.12)$$

e. Reliability Function, Hazard Function, and MTTF:

$$R(x) = \exp\left[-(x)^c\right]; x > 0, c > 0 \tag{5.2.2.13}$$

$$\lambda(x) = c(x)^{c-1} \quad ; x > 0, c > 0 \tag{5.2.2.14}$$

$$\text{MTTF} = \Gamma(1 + (1/c)) \tag{5.2.2.15}$$

f. Moment Generating Function (MGF), Characteristic Function (CF) and Entropy:

$$M_X(t) = \sum_{r=0}^{\infty} \frac{(t)^r}{r!} \Gamma\left(1 + (r/c)\right) \tag{5.2.2.16}$$

$$\phi_X(t) = \sum_{r=0}^{\infty} \frac{(it)^r}{r!} \Gamma\left(1 + (r/c)\right) \tag{5.2.2.17}$$

$$\text{Entropy} = \log_e \frac{1}{c} + (c-1)\frac{\Upsilon}{c} + 1 \tag{5.2.2.18}$$

5.3 RAYLEIGH DISTRIBUTION

It is a special case of the Weibull model. Rayleigh model is named after the British physicist Lord Rayleigh (1842–1919). This model has wide applications in modern science and technology, specifically, communication theory, signal processing, aircraft technology, reliability, and survival analysis, etc.

5.3.1 TWO-PARAMETER RAYLEIGH DISTRIBUTION

Two-parameters Rayleigh model can be derived by putting c = 2 in eq (5.1.1) and (5.1.2). The CDF and PDF of this model are respectively given by

$$F_X(x) = 1 - \exp\left[-\left(\frac{x-a}{b}\right)^2\right]; \qquad x > a \tag{5.3.1.1}$$

Symbolically: $X \sim R(a,b)$

$$f_X(x) = 2\frac{1}{b}\left(\frac{x-a}{b}\right)\exp\left[-\left(\frac{x-a}{b}\right)^2\right] \; ; x > a,\ b > 0,\ -\infty < a < \infty \tag{5.3.1.2}$$

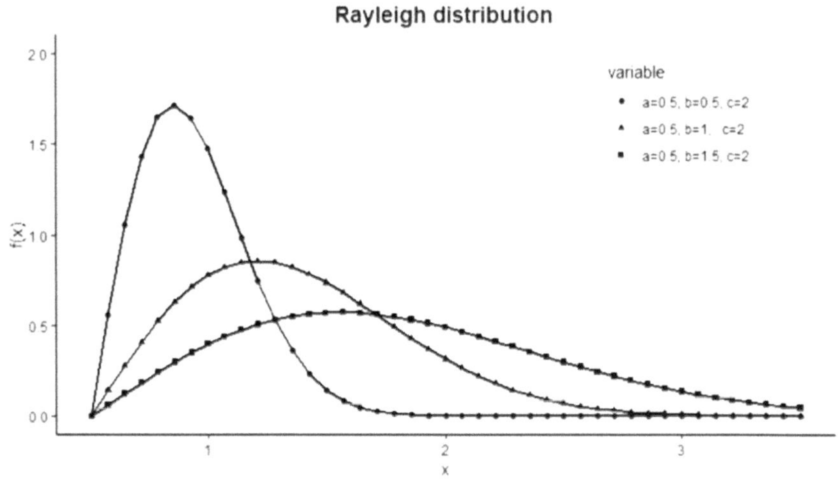

FIGURE 5.4 Rayleigh distribution.

Sample Generation of Rayleigh Distribution

R Code: Generate random sample from Rayleigh Distribution.

x = rweibull(n, 2, b)

where; n is the sample size; b is the parameters of the distribution

Example:

x = rweibull(100, 2, 1)

```
Console  Terminal  Jobs
~/
> x= rweibull(10, 2, 1)
> x
 [1] 1.4851130 1.7085012 1.2708991 1.0704167 0.3934033 0.6748788 0.2875594 0.5591951 0.8529504 1.0076963
```

If $a \neq 0$

- **Step 1:** Generate x= rweibull(n, c, b)
- **Step 2:** Generate $y = x + a$

i.e., y is a random sample size n from Rayleigh (a, b).

The properties of two-parameter Rayleigh Distribution are given below

5.3.1.1 PROPERTIES OF TWO-PARAMETER RAYLEIGH DISTRIBUTION

a. Moments of Two-Parameter Rayleigh Distribution:

$$\mu_1' = \frac{1}{2} b\sqrt{\pi} + a \tag{5.3.1.3}$$

$$\mu_2' = b^2 + a^2 + ab\sqrt{\pi} \tag{5.3.1.4}$$

$$\mu_3' = \frac{3}{4} b^3\sqrt{\pi} + a^3 + \frac{3}{2} a^2 b\sqrt{\pi} + 3ab^2 \tag{5.3.1.5}$$

$$\mu_4' = 2b^4 + a^4 + 2a^3 b\sqrt{\pi} + 3ab^3\sqrt{\pi} + 6a^2b^2 \tag{5.3.1.6}$$

b. Central Moments of Two-Parameter Rayleigh Distribution:

$$\mu_1 = 0 \tag{5.3.1.7}$$

$$\mu_2 = \frac{1}{4} b^2 [4-\pi] \tag{5.3.1.8}$$

$$\mu_3 = \frac{1}{8} b^3\sqrt{\pi}\left[6 - 12\sqrt{\pi} + \pi\right] \tag{5.3.1.9}$$

$$\mu_4 = \frac{1}{8} b^4\left(16 - 3\pi^2\right) \tag{5.3.1.10}$$

c. Skewness and Kurtosis of Two-Parameter Rayleigh Distribution:

$$\beta_1 = \frac{\pi\left[6 - 12\sqrt{\pi} + \pi\right]^2}{[4-\pi]^3} \tag{5.3.1.11}$$

$$\gamma_1 = \frac{\sqrt{\pi}\left[6 - 12\sqrt{\pi} + \pi\right]}{[4-\pi]^{3/2}} \tag{5.3.1.12}$$

$$\beta_2 = \frac{\left(16 - 3\pi^2\right)}{[4-\pi]^2} \tag{5.3.1.13}$$

$$\gamma_2 = \frac{4 - 6\pi^2 + 24\pi}{[4-\pi]^2} \tag{5.3.1.14}$$

d. Median, Quartile, Percentile, and Mode of Two-Parameter Rayleigh Distribution:

$$\text{Median} = b\left(\log_e 2\right)^{1/2} + a \tag{5.3.1.15}$$

$$Q_r = b\left(-\log_e \frac{r}{4}\right)^{1/2} + a \quad ;r = 1,2,3$$

$$P_r = b\left(-\log_e \frac{r}{100}\right)^{1/2} + a \quad ;r = 1,2,\ldots,99$$

$$\text{Mode} = b\sqrt{\left(\frac{1}{2}\right)} + a \tag{5.3.1.16}$$

e. Reliability Function, Hazard Function, and MTTF of Two-Parameter Rayleigh Distribution

$$R(x) = \exp\left[-\left(\frac{x-a}{b}\right)^2\right]; x > a \tag{5.3.1.17}$$

$$\lambda(x) = 2\frac{1}{b}\left(\frac{x-a}{b}\right) \; ; x > a \tag{5.3.1.18}$$

$$\text{MTTF} = \frac{1}{2} b\sqrt{\pi} + a \tag{5.3.1.19}$$

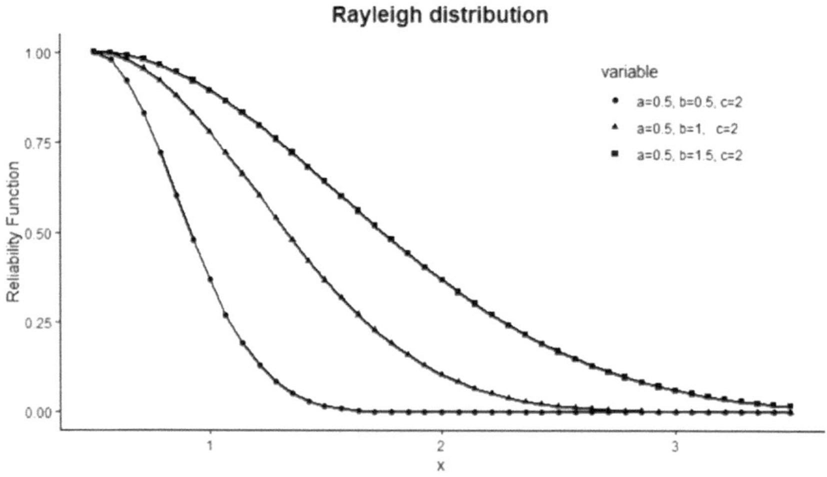

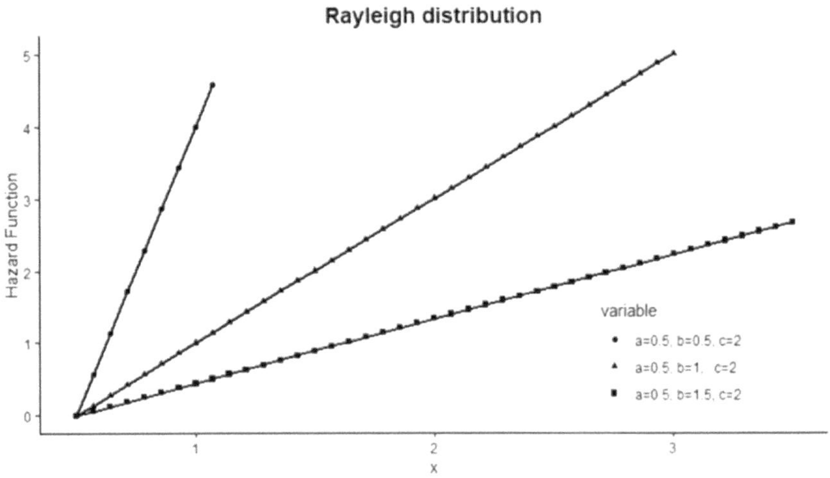

FIGURE 5.5 Reliability function and hazard function of Rayleigh distribution.

f. Moments Generating Function, Characteristic Function (CF), and Entropy of Two Parameter Rayleigh Distribution

$$M_X(t) = e^{ta} \sum_{r=0}^{\infty} \frac{(tb)^r}{r!} \Gamma\left(1+(r/2)\right) \tag{5.3.1.20}$$

$$\phi_X(t) = e^{ita} \sum_{r=0}^{\infty} \frac{(itb)^r}{r!} \Gamma\left(1+(r/2)\right) \tag{5.3.1.21}$$

$$\text{Entropy} = \log_e \frac{b}{2} + \frac{\Upsilon}{2} + 1 \tag{5.3.1.22}$$

5.3.2 ONE PARAMETER RAYLEIGH DISTRIBUTION

One parameter Rayleigh model can be derived by putting c = 2 and a = 0 in eq (5.1.1) and (5.1.2). The CDF and PDF of this model are respectively given by

$$F_X(x) = 1 - \exp\left[-\left(\frac{x}{b}\right)^2\right]; \qquad x > 0, b > 0 \tag{5.3.2.1}$$

Symbolically: $X \sim R(b)$

$$f_X(x) = 2\frac{1}{b}\left(\frac{x}{b}\right)\exp\left[-\left(\frac{x}{b}\right)^2\right]; \; x > 0, \; b > 0 \tag{5.3.2.2}$$

The properties of one parameter Rayleigh Distribution are given below

5.3.2.1 PROPERTIES OF ONE PARAMETER RAYLEIGH DISTRIBUTION

a. Moments of One Parameter Rayleigh Distribution:

$$\mu_1' = \frac{1}{2} b\sqrt{\pi} \tag{5.3.2.3}$$

$$\mu_2' = b^2 \tag{5.3.2.4}$$

$$\mu_3' = \frac{3}{4} b^3 \sqrt{\pi} \tag{5.3.2.5}$$

$$\mu_4' = 2b^4 \tag{5.3.2.6}$$

b. The first four Central Moments of one parameter Rayleigh model are same as two-parameter Rayleigh model (given in Eqns. (5.3.1.7)–(5.3.1.10)).

c. Skewness and Kurtosis of one parameter Rayleigh model are same as two-parameter Rayleigh model (given in Eqns. (5.3.1.11)–(5.3.1.14)).

d. Median, Quartile, Percentile, and Mode of One Parameter Rayleigh Distribution

$$\text{Median} = b(\log_e 2)^{1/2} \tag{5.3.2.7}$$

$$Q_r = b\left(-\log_e \frac{r}{4}\right)^{1/c} \quad ; r = 1,2,3$$

$$P_r = b\left(-\log_e \frac{r}{100}\right)^{1/c} \quad ; r = 1,2,...,99$$

$$\text{Mode} = b\sqrt{\left(\frac{1}{2}\right)} \tag{5.3.2.8}$$

e. Reliability Function, Hazard Function, and MTTF of One Parameter Rayleigh Distribution

$$R(x) = \exp\left[-\left(\frac{x}{b}\right)^2\right] ; x > 0 \tag{5.3.2.9}$$

$$\lambda(x) = 2\frac{1}{b}\left(\frac{x}{b}\right) ; x > 0 \tag{5.3.2.10}$$

$$\text{MTTF} = b\overline{\left|3/2\right.} = \frac{1}{2} b\sqrt{\pi} \tag{5.3.2.11}$$

f. Moments Generating Function, Characteristic Function (CF), and Entropy of One Parameter Rayleigh Distribution

$$M_X(t) = \sum_{r=0}^{\infty} \frac{(tb)^r}{r!} \overline{\left|1+(r/2)\right.} \tag{5.3.2.12}$$

$$\phi_X(t) = \sum_{r=0}^{\infty} \frac{(itb)^r}{r!} \overline{\left|1+(r/2)\right.} \tag{5.3.2.13}$$

$$\text{Entropy} = \log_e \frac{b}{2} + \frac{\Upsilon}{2} + 1 \tag{5.3.1.14}$$

5.4 EXPONENTIAL DISTRIBUTION

The exponential distribution is a popular distribution in industrial applications, where the units manufactured using the standard process are modeled by two parameters exponential distribution. It also arises naturally when describing the waiting time in a homogeneous Poisson Process. It is also a useful model for queuing theory, reliability theory, physics, and hydrology, etc. It is a submodel of the Weibull model.

5.4.1 TWO-PARAMETER EXPONENTIAL DISTRIBUTION

Two-parameters exponential model can be derived by putting c = 1 in eq (5.1.1) and (5.1.2). The CDF and PDF of this model are respectively given by

$$F_X(x) = 1 - \exp\left[-\left(\frac{x-a}{b}\right)\right]; \qquad x > a, b > 0 \tag{5.4.1.1}$$

Symbolically: $X \sim Exp(a,b)$

$$f_X(x) = \frac{1}{b}\exp\left[-\left(\frac{x-a}{b}\right)\right]; \; x > a,\ b > 0,\ -\infty < a < \infty \tag{5.4.1.2}$$

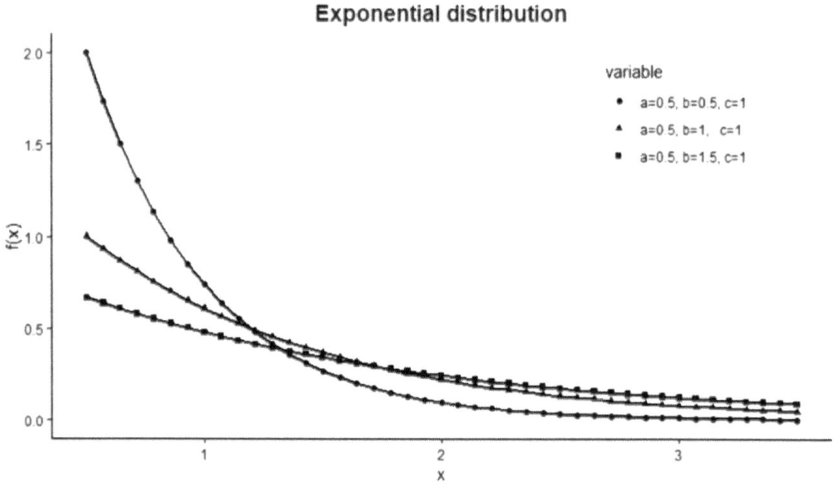

FIGURE 5.6 Exponential distribution.

Sample Generation of Exponential Distribution

R Code: Generate random sample from Exponential Distribution.

x = rexp(n, d)

where; n is the sample size; d is the reciprocal of the scale parameter

Example:

x = rexp(100, 1)

If $a \neq 0$

- **Step 1:** Generate x = rexp(n, d)
- **Step 2:** Generate $y = x + a$

i.e., y denotes the vector containing Exponential (a,d) data set of size n.

5.4.1.1 PROPERTIES OF TWO-PARAMETER EXPONENTIAL DISTRIBUTION

a. Moments of Two-Parameter Exponential Distribution

$$\mu_1' = b + a \tag{5.4.1.3}$$

$$\mu_2' = 2b^2 + a^2 + 2ab \tag{5.4.1.4}$$

$$\mu_3' = 6b^3 + a^3 + 6a^2b + 6ab^2 \quad (5.4.1.5)$$

$$\mu_4' = 24b^4 + a^4 + 4a^3b + 24ab^3 + 12a^2b^2 \quad (5.4.1.6)$$

b. Central Moments of Two-Parameter Exponential Distribution

$$\mu_1 = 0 \quad (5.4.1.7)$$

$$\mu_2 = b^2 \quad (5.4.1.8)$$

$$\mu_3 = 2b^3 \quad (5.4.1.9)$$

$$\mu_4 = 9b^4 \quad (5.4.1.10)$$

c. Skewness and Kurtosis of Two-Parameter Exponential Distribution

$$\beta_1 = 4 \quad (5.4.1.11)$$

$$\gamma_1 = 2 \quad (5.4.1.12)$$

$$\beta_2 = 9 \quad (5.4.1.13)$$

$$\gamma_2 = 6 \quad (5.4.1.14)$$

d. Median, Quartile, Percentile, and Mode of the Two-Parameter Exponential Distribution

$$\text{Median} = b(\log_e 2) + a \quad (5.4.1.15)$$

$$Q_r = b\left(-\log_e \frac{r}{4}\right) + a \quad ;r = 1,2,3$$

$$P_r = b\left(-\log_e \frac{r}{100}\right) + a \quad ;r = 1,2,\ldots,99$$

$$\text{Mode} = a \quad (5.4.1.16)$$

e. Reliability Function, Hazard Function, MTTF of Two-Parameter Exponential Distribution

$$R(x) = \exp\left[-\left(\frac{x-a}{b}\right)\right]; x > a \quad (5.4.1.17)$$

$$\lambda(x) = \frac{1}{b} \quad (5.4.1.18)$$

Constant hazard rate.

$$\text{MTTF} = b + a \quad (5.4.1.19)$$

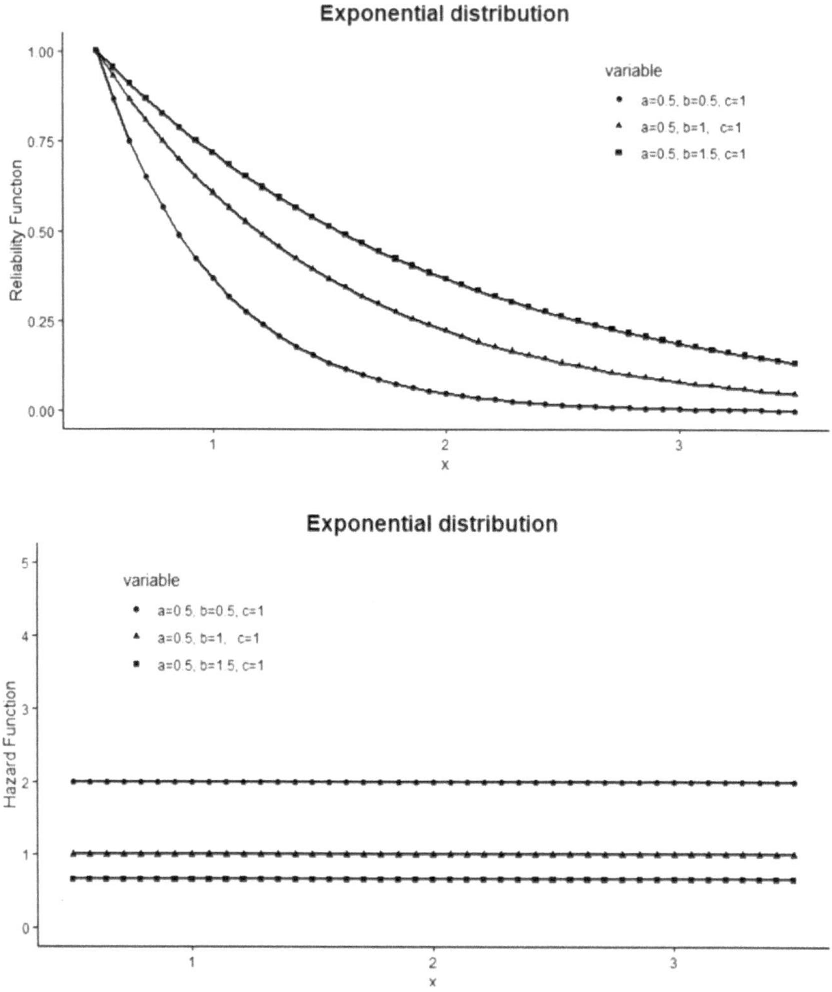

FIGURE 5.7 Reliability function and hazard function of Rayleigh distribution.

f. Moments Generating Function and Characteristic Function (CF) of Two-Parameter Exponential Distribution

$$M_X(t) = e^{ta}\left(1 - tb\right)^{-1}; t < \frac{1}{b} \tag{5.4.1.20}$$

$$\phi_X(t) = e^{ita}\left(1 - itb\right)^{-1}; t < \frac{1}{b} \tag{5.4.1.21}$$

5.4.2 ONE PARAMETER EXPONENTIAL DISTRIBUTION

One parameter Exponential model can be derived by putting c = 1 and a = 0 in eq (5.1.1) and (5.1.2). The CDF and PDF of this model are respectively given by

$$F_X(x) = 1 - \exp\left[-\left(\frac{x}{b}\right)\right]; \qquad x > 0, b > 0 \qquad (5.4.2.1)$$

Symbolically: X ~ *Exp*(*b*)

$$f_X(x) = \frac{1}{b}\exp\left[-\left(\frac{x}{b}\right)\right]; \; x > 0, \; b > 0 \qquad (5.4.2.2)$$

5.4.2.1 PROPERTIES OF ONE PARAMETER EXPONENTIAL DISTRIBUTION

a. Moments:

$$\mu_1' = b \qquad (5.4.2.3)$$

$$\mu_2' = 2b^2 \qquad (5.4.2.4)$$

$$\mu_3' = 6b^3 \qquad (5.4.2.5)$$

$$\mu_4' = 24b^4 \qquad (5.4.2.6)$$

b. The first four Central Moments of one parameter exponential model are same as two-parameter exponential model (given in Eqns. (5.4.1.7) –(5.4.1.10))

c. Skewness and Kurtosis of one parameter exponential model are same as two-parameter exponential model (given in Eqns. (5.4.1.11) –(5.4.1.14))

d. Median, Quartile, Percentile, and Mode

$$\text{Median} = b(\log_e 2) \qquad (5.4.2.7)$$

$$Q_r = b\left(-\log_e \frac{r}{4}\right) \qquad ;r=1,2,3$$

$$P_r = b\left(-\log_e \frac{r}{100}\right) \qquad ;r=1,2,...,99$$

$$\text{Mode} = 0 \tag{5.4.2.8}$$

e. Reliability Function, Hazard Function, MTTF

$$R(x) = \exp\left[-\left(\frac{x}{b}\right)\right]; x > 0 \tag{5.4.2.9}$$

$$\lambda(x) = \frac{1}{b}. \tag{5.4.2.10}$$

It is a constant hazard rate and it is the same as two parameter exponential distribution.

$$\text{MTTF} = b\lceil 2 = b \tag{5.4.2.11}$$

f. Moment Generating Function (MGF) and Characteristic Function (CF)

$$M_X(t) = (1 - tb)^{-1}; t < \frac{1}{b} \tag{5.4.2.12}$$

$$\phi_X(t) = (1 - itb)^{-1}; t < \frac{1}{b} \tag{5.4.2.13}$$

g. Cumulant Generating Function (CGF): The expression for CGF $(K_X(t))$ is obtained as below:

$$K_X(t) = \log(M_X(t))$$

$$K_X(t) = \log(1 - tb)^{-1} \quad \text{[using Eqns. (5.4.2.12)]} \tag{5.4.2.14}$$

We can get central moments using the cumulants generating function using two methods.

In these two methods, we can find all fours κ_1, κ_2, κ_3, κ_4 using:

i. κ_1 = coefficient of $\frac{t^r}{r!}$ in $K_X(t)$

ii. $\kappa_r = \frac{\partial^r}{\partial t^r} K_X(t)\Big| t = 0.$

κ_1 = mean; κ_2 = variance; $\kappa_3 = \mu_3$; $\kappa_4 = -3\kappa_2^2 + \mu_4$.

h. Memoryless Property: This is the most important property of exponential distribution. If X is exponential with parameter b, then the memoryless property can be defined as

$$P(X > d + x/X > d) = P(X > x); \quad \text{for d,x} \geq 0$$

$$P(X > d + x / X > a) = \frac{P(X > d + x, X > d)}{P(X > d)}; \quad \text{for d,x} \geq 0$$

$$= \frac{P(X > d + x)}{P(X > d)} = \frac{1 - F_X(d + x)}{1 - F_X(d)}$$

$$= \frac{e^{-b(x+d)}}{e^{-bd}} = e^{-bx} = P(X > x)$$

Therefore,

$$P(X > d + x / X > d) = P(X > x); \quad \text{for d,x} \geq 0$$

5.5 SOME MODIFICATIONS OF WEIBULL DISTRIBUTION

Although, the Weibull model is one of the most useful models for analyzing the lifetime/failure time data, the data sets with bathtub shape hazard rates cannot be analyzed by the Weibull model. For analyzing these data sets, some modified Weibull distributions are available in the statistical literature. Now, we will discuss a few modified Weibull distributions.

5.5.1 *EXPONENTIATED WEIBULL DISTRIBUTION*

Mudholkar and Srivastava (1993) developed a three-parameter exponentiated Weibull model by adding a new shape parameter $d > 0$ in the standard two-parameter Weibull model. The PDF and CDF of this model is:

$$f_X(x) = \frac{cdx^{c-1}}{b^c} e^{-\left(\frac{x}{b}\right)^c} \left[1 - e^{-\left(\frac{x}{b}\right)^c}\right]^{d-1} \quad ; x > 0$$

and $F_X(x) = \left[1 - e^{-\left(\frac{x}{b}\right)^c}\right]^d \quad ; x > 0$

where $b > 0$ denotes scale parameter and $c > 0$, $d > 0$ denote shape parameters. Mudholkar et al. (1995) further studied various properties of exponentiated Weibull distribution and analyzed the bus-motor failure data

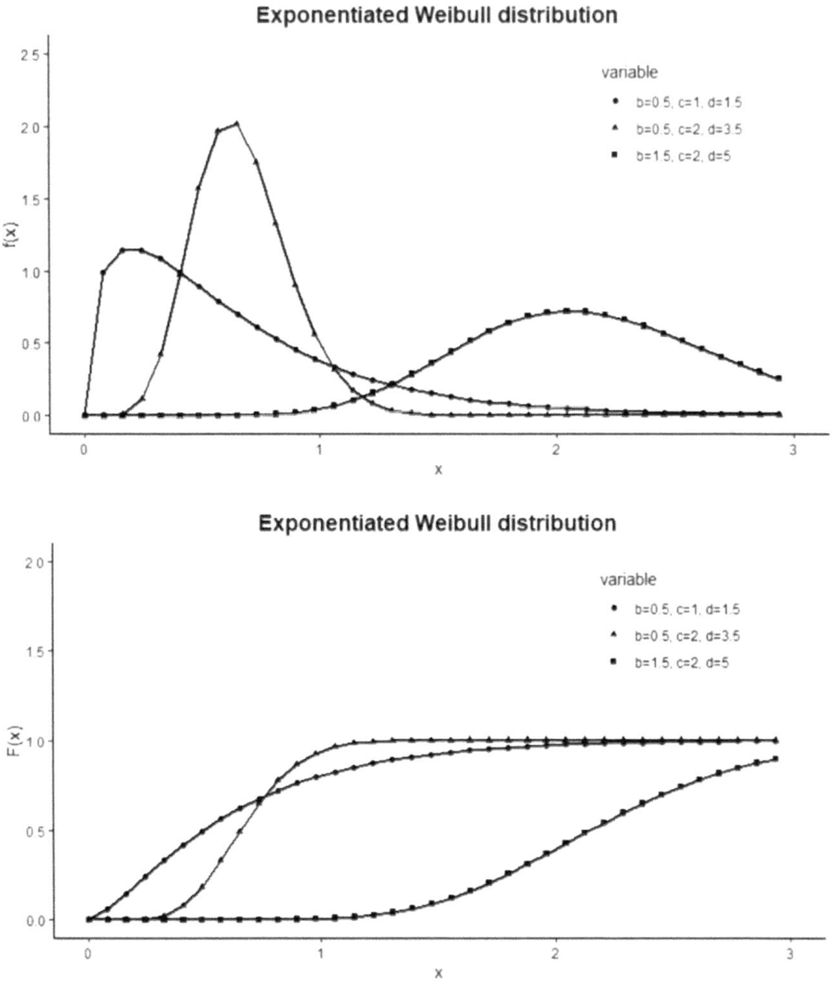

FIGURE 5.8 Density function and distribution function of exponentiated Weibull distribution.

The particular cases of this model are given below:

Parameters	Name of Distribution
$d = 1$	Two-parameter Weibull distribution
$c = 2$	Generalized Rayleigh distribution
$c = 1$	Generalized exponential distribution (see Kundu and Gupta, 1999)
$d = 1, c = 2$	Rayleigh distribution
$d = 1, c = 1$	Exponential distribution

The hazard function of the distribution is given by

$$\lambda(x) = \frac{1}{R(x)} f(x)$$

$$= \frac{1}{\left[1-\left[1-e^{-\left(\frac{x}{b}\right)^c}\right]^d\right]} \frac{cdx^{c-1}}{b^c} e^{-\left(\frac{x}{b}\right)^c} \left[1-e^{-\left(\frac{x}{b}\right)^c}\right]^{d-1}, x>0$$

The following table provides the failure rate behavior of this distribution (see page 300, Mudholkar and Srivastava, 1993).

c	d	**Failure Rate Behavior**
1	1	Constant (exp. distribution)
	1	Monotonic (Weibull distribution)
<1	<1	Decreasing
>1	>1	Increasing
>1	<1	Bathtub or increasing
<1	>1	Unimodal or decreasing

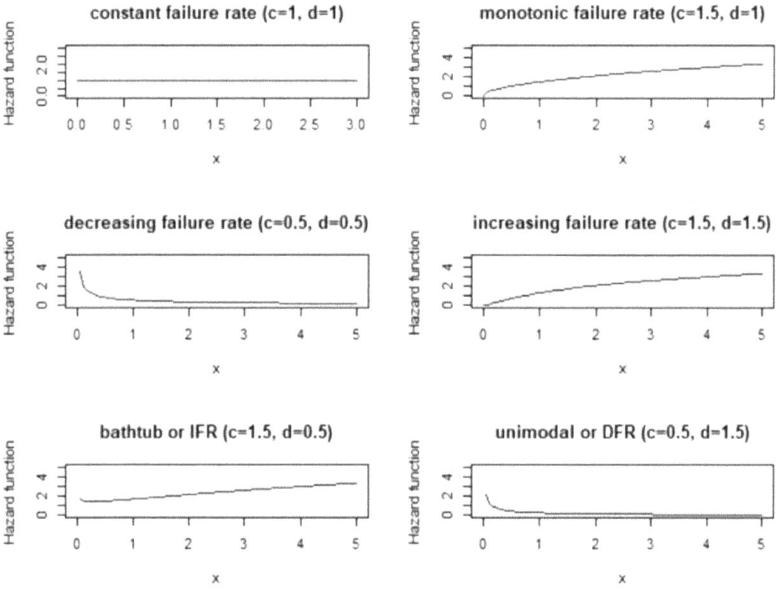

FIGURE 5.9 Failure rate.

5.5.2 INVERSE WEIBULL DISTRIBUTION

If X follows a Weibull distribution, then $Y = \frac{1}{X}$ has the inverse Weibull distribution. The CDF, PDF, and hazard function of inverse Weibull distribution is respectively, given by

$$F_Y(y) = e^{-\left(\frac{1}{by}\right)^c} ; y > 0, b > 0, c > 0$$

$$f_Y(y) = c\frac{1}{b^c y^{c+1}} e^{-\left(\frac{1}{by}\right)^c} ; y > 0, b > 0, c > 0$$

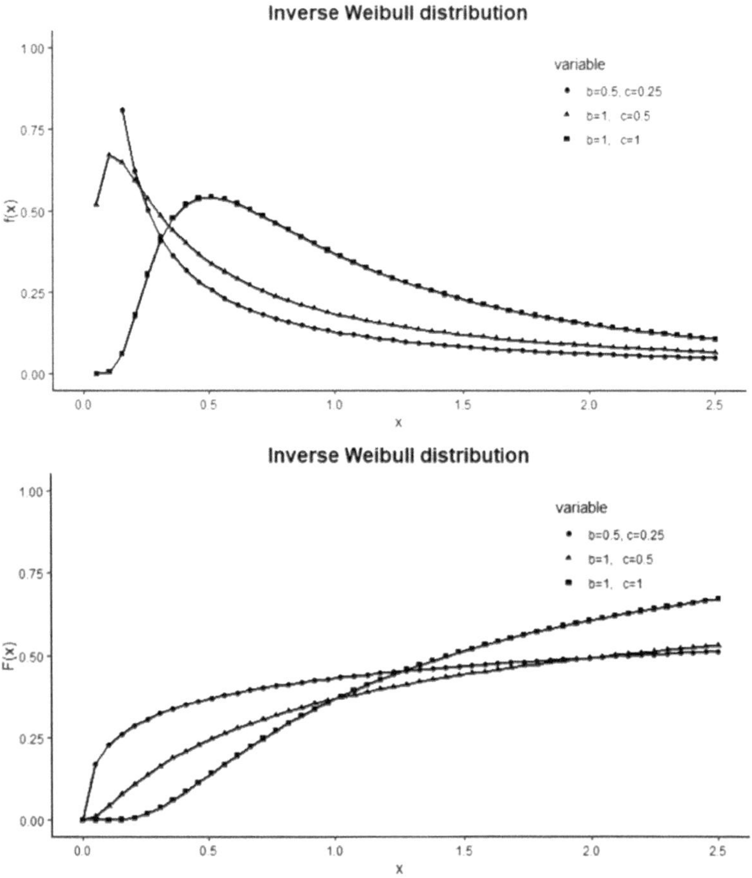

FIGURE 5.10 Density function and distribution function of inverse Weibull distribution.

$$\lambda(y) = \frac{1}{\left(1 - e^{-\left(\frac{1}{by}\right)^c}\right)} c \frac{1}{b^c y^{c+1}} e^{-\left(\frac{1}{by}\right)^c}$$

$$= c \frac{1}{b^c y^{c+1}} \frac{1}{\left(e^{\left(\frac{1}{by}\right)^c} - 1\right)}; y > 0, b > 0, c > 0$$

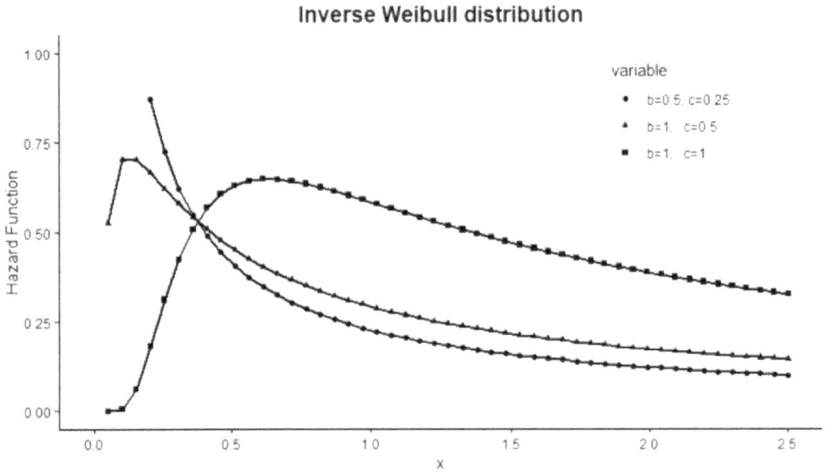

FIGURE 5.11 Hazard function of inverse Weibull distribution.

5.5.3 MODIFIED WEIBULL DISTRIBUTION

Gurvich et al. (1997) developed a modified Weibull distribution in the context of modeling random strength of brittle materials. Lai et al. (2003) independently proposed the same distribution for modeling failure times. The authors studied the several properties of this model. The CDF and PDF of this model are, respectively, given by

$$F_X(x) = 1 - \exp\left[-\left(\frac{x}{b}\right)^c \exp(dx)\right]; x > 0 \qquad (5.5.1)$$

$$f_X(x) = \left[\frac{dx+c}{b}\right]\left(\frac{x}{b}\right)^{c-1} \exp(ax)\exp\left[-\left(\frac{x}{b}\right)^{c} \exp(dx)\right]; x>0\ ;\ d>0,\ b>0,\ c>0$$

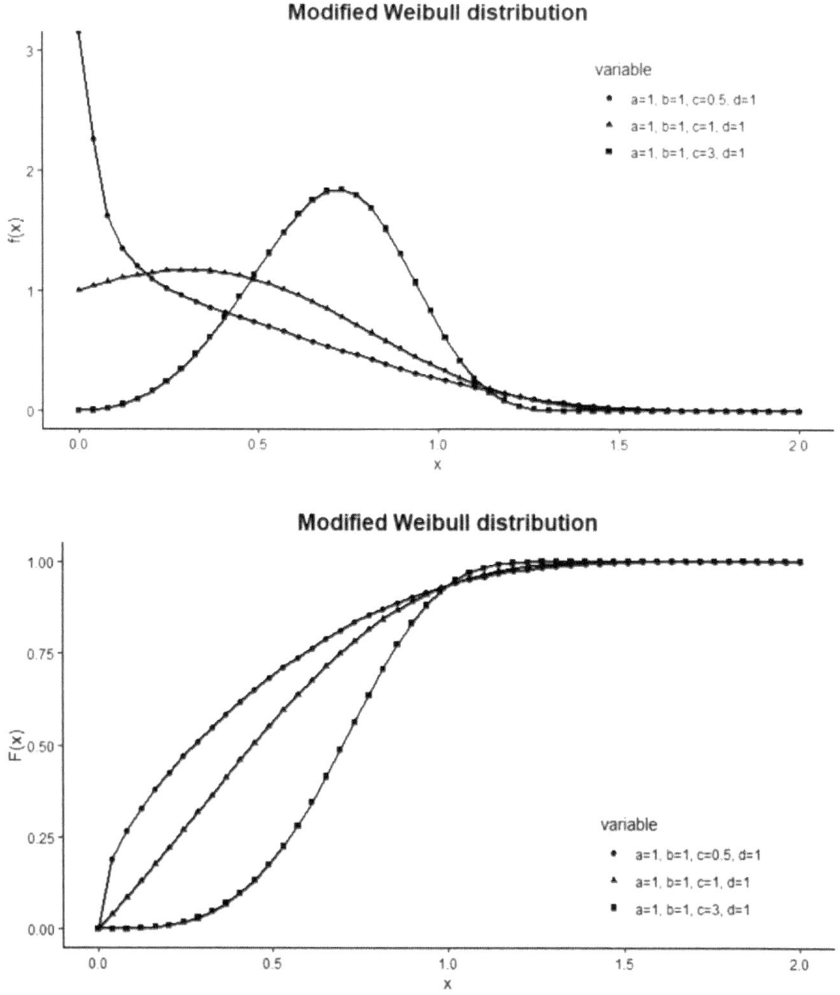

FIGURE 5.12 Density function and distribution function of modified Weibull distribution.

The hazard function is given by

$$\lambda(x) = \left[\frac{dx+b}{c}\right]\left(\frac{x}{b}\right)^{c-1} \exp(dx); x>0$$

Since the parameters b and d do not influence the slope of $\lambda(x)$, we may consider the two cases, namely, $c \in (0,1)$ and $c \geq 1$. In the case $c \in (0,1)$, $\lambda(x)$ initially decreases and then increases in x, i.e., $\lambda(x)$ has a bathtub shape. Recall that the hazard function of the usual Weibull distribution is constant, increasing, or decreasing. Thus, the hazard function $\lambda(x)$ for the modified Weibull distribution differs from the usual Weibull distribution (Figure 5.13). In another case $c \geq 1$, the hazard function $\lambda(x)$ in increasing function of x, i.e., the modified Weibull model has an increasing hazard function. For more details about the modified Weibull model, the readers are advised to see Gurvich et al. (1997); and Lai et al. (2003).

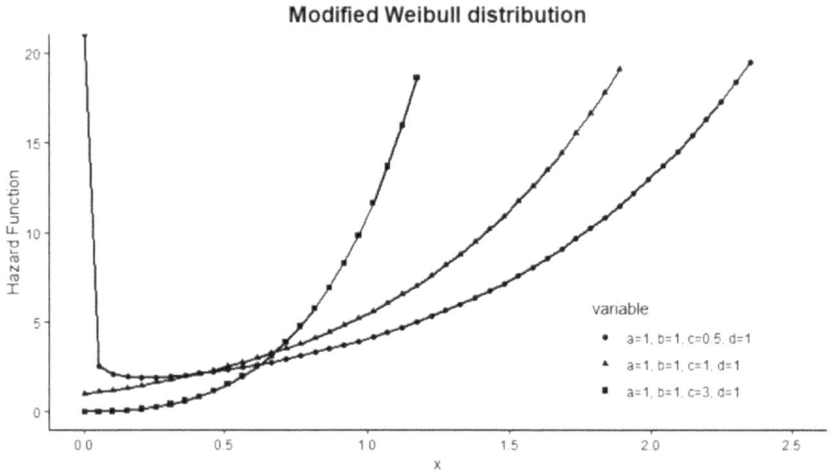

FIGURE 5.13 Hazard function of modified Weibull distribution.

KEYWORDS

- **Characteristic function**
- **Cumulative distribution function**
- **Meantime to failure**
- **Moment generating function**
- **Probability density function**

REFERENCES

Fréchet, M., *(1927). On the Law of Probability of the Maximum Deviation* (Vol. *6, pp. 93–116*). Annals of the Polish Mathematical Society, Krakow.

Gurvich, M. R., Dibenedetto, A. T., & Rande, S. V., (1997). A new statistical distribution for characterizing the random strength of brittle materials. *Journal of Materials Science, 32*, 2559–2564.

Kundu, D., & Gupta, R. D., (1999). Generalized exponential distribution. *Aust. N Z J. Stat., 41*, 173–188.

Lai, C. D., Xie, M., & Murthy, D. N. P., (2003). A modified Weibull distribution. *IEEE Transactions on Reliability, 52*(1), 33–37.

Mudholkar, G. S., & Srivastava, D. K., (1993). Exponentiated Weibull family for analyzing bathtub failure rate data. *IEEE Transactions on Reliability, 42*(2), 299–302.

Mudholkar, G. S., Srivastava, D. K., & Freimer, M., (1995). The exponentiated Weibull family: A reanalysis of the bus-motor-failure data. *Technometrics, 37*(4), 436–445.

Murthy, D. N. P., Xie, M., & Jiang, R., (2003). *Weibull Models* (Vol. 358). New York: Wiley.

Rinne, H., (2009). *The Weibull Distribution: A Handbook.* CRC Press, Taylor & Francis.

Rosin, P., & Rammler, E., (1933). The laws governing the fineness of powdered coal. *Journal of the Institute of Fuel, 7*, 29–36.

Shannon, C. E., (1948). The mathematical theory of communication. *Bell Syst. Tech. J., 27*, 249–428.

Weibull, W., (1951). A statistical distribution function of wide applicability. *Journal of Applied Mechanics, 18,* 293–297.

Weibull, W., (1952). Discussion: A statistical distribution function of wide applicability. *Journal of Applied Mechanics, 19,* 233–234.

Weibull, W., (1961). *Fatigue Testing and Analysis of Results.* Pergamon Press, New York.

Xie, M., Tang, Y., & Goh, T. N., (2002). A modified Weibull extension with bathtub-shaped failure rate function. *Reliability Engineering System Safety, 76*(3), 279–285.

CHAPTER 6

Life Distributions

MOHD. ARSHAD[1], MUKTI KHETAN[2], VIJAY KUMAR[3], and FOZIA HOMA[4]

[1]*Department of Mathematics, Indian Institute of Technology Indore, Simrol, Indore, Madhya Pradesh, India*

[2]*Department of Mathematics, Amity School of Applied Sciences, Amity University Mumbai, Maharashtra, India*

[3]*Department of Mathematics, Indian Institute of Technology Bombay, Powai, Mumbai, Maharashtra, India*

[4]*Department of Statistics, Mathematics, and Computer Application, Bihar Agricultural University, Sabour, Bhagalpur, Bihar, India*

6.1 PARETO DISTRIBUTION

The Pareto distribution is named after an Italian-born Swiss professor of economics, civil engineer, and sociologist, Vilfredo Pareto (1897). Pareto developed a law, which dealt with the distribution of income or distribution of wealth over a population or a society, known as Pareto's Law. His original quotation towards the distribution of income or wealth explained it as an 80–20 proportion. In simple terms, the smaller proportion of the population experiencing the larger portion of wealth. This 80–20 rule is also known as the Pareto principle.

Distribution Theory: Principles and Applications. Fozia Homa, Mukti Khetan, Mohd. Arshad, & Pradeep Mishra (Naz) (Eds.)

➢ **Definition:** If a random variable X follows the Pareto distribution, then the expression of the cumulative distribution function (CDF) is

$$F_X(x) = 1 - \left(\frac{\lambda}{x}\right)^{\alpha}; \qquad \lambda < x < \infty$$

Symbolically: $X \sim P(\alpha, \lambda)$

Here α, λ (>0) are the parameters of the distribution.

The probability density function (PDF) is

$$f_X(x) = \frac{d}{dx} F_X(x) = \frac{d}{dx}\left\{1 - \left(\frac{\lambda}{x}\right)^{\alpha}\right\}$$

$$f_X(x) = \frac{\alpha\lambda^{\alpha}}{x^{\alpha+1}}\ ; \ x > \lambda,\ \alpha > 0,\ \lambda > 0.$$

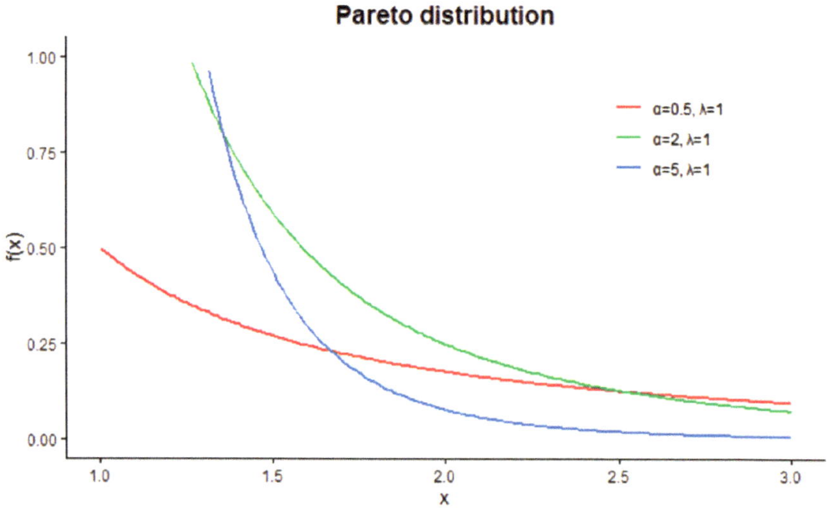

FIGURE 6.1 Pareto distribution.

Sample Generation of Using Inverse Transformation Method:

$$F_X(x) = 1 - \left(\frac{\lambda}{x}\right)^{\alpha}; \qquad x > \lambda$$

$$U = 1 - \left(\frac{\lambda}{x}\right)^{\alpha}$$

$$(1-U)=\left(\frac{\lambda}{x}\right)^{\alpha}$$

$$X=\frac{\lambda}{(1-U)^{1/\alpha}},$$

where U follows uniform (0,1).

- **Note:** Firstly, we generate a random sample of size n from $U(0,1)$, i.e., $\underline{u}$ = $(u_1, u_2, \ldots, u_n)$ (say). Define, $x_i=\frac{\lambda}{(1-u_i)^{1/\alpha}};\ i=1,2,..,n$. Then, $\underline{x}$ = $(x_1, x_2, \ldots, x_n)$ gives a random sample of size n from $P(\alpha, \lambda)$.
- **Example 6.1:** Let u = (0.1,0.3,0.22,0.51,0.11,0.6) be a random sample generated from $U(0,1)$ distribution. Now, we want to generate data from $P(2,3)$, $P(2,5)$ and $P(3,5)$. Calculate $\frac{\lambda}{(1-u_i)^{1/\alpha}};\ i=1,2,..,6,$ where u_i denotes the ith element of = $(u_1, u_2, \ldots, u_n)$

***U*(0,1)**	**Random Sample**		
	***P*(2,5)**	***P*(2,5)**	***P*(3,5)**
0.10	3.162278	5.270463	5.178721
0.30	3.585686	5.976143	5.631239
0.22	3.396831	5.661385	5.431734
0.51	4.285714	7.142857	6.342171
0.11	3.179994	5.299989	5.198045
0.60	4.743416	7.905694	6.786044

6.1.1 PROPERTIES OF THE PARETO DISTRIBUTION

a. **Raw Moments of Pareto Distribution:** The first four raw moments $\mu_1', \mu_2', \mu_3', \mu_4'$ have been derived below.

   r^th^ raw moment

$$\mu_r' = E\left(X^r\right)$$

$$=\int_{\lambda}^{\infty} x^r \alpha \frac{\lambda^{\alpha}}{x^{\alpha+1}}\,dx$$

$$=\alpha\lambda^{\alpha}\int_{\lambda}^{\infty} x^{-\alpha+r-1}dx$$

$$= \alpha\lambda^{\alpha}\left[\frac{x^{-\alpha+r}}{-\alpha+r}\right]_{\lambda}^{\infty} = \alpha\lambda^{\alpha}\frac{\lambda^{-\alpha+r}}{\alpha-r}$$

$$\mu'_r = \frac{\lambda^r\alpha}{(\alpha-r)}; \quad \alpha > r \quad ; r = 1,2,3,4$$

$$\text{For } r = 1,\ \mu'_1 = \frac{\lambda\alpha}{(\alpha-1)}; \quad \alpha > 1, \tag{6.1.1}$$

$$\text{For } r = 2,\ \mu'_2 = \frac{\lambda^2\alpha}{(\alpha-2)}; \quad \alpha > 2, \tag{6.1.2}$$

$$\text{For } r = 3,\ \mu'_3 = \frac{\lambda^3\alpha}{(\alpha-3)}; \quad \alpha > 3, \tag{6.1.3}$$

$$\text{For } r = 4,\ \mu'_4 = \frac{\lambda^4\alpha}{(\alpha-4)}; \quad \alpha > 4 \tag{6.1.4}$$

b. **Central Moments of Pareto Distribution:** The first four central moments μ_1, μ_2, μ_3, μ_4 have been derived in this section.

We have already discussed that the first central moment is always zero

$$\mu_1 = 0 \tag{6.1.5}$$

For the second central moment, we are utilizing the values of the first two raw moments and we get:

$$\mu_2 = \mu'_2 - \mu'^2_1$$

$$= \frac{\lambda^2\alpha}{(\alpha-2)} - \left[\frac{\lambda\alpha}{(\alpha-1)}\right]^2 \quad \text{[using Eqns. (6.1.1) and (6.1.2)]}$$

$$= \frac{\lambda^2\alpha}{(\alpha-2)(\alpha-1)^2}\left[(\alpha-1)^2 - (\alpha-2)\alpha\right]$$

$$= \frac{\lambda^2\alpha}{(\alpha-2)(\alpha-1)^2}\left[\left(1+\alpha^2-2\alpha\right)-\left(-2\alpha+\alpha^2\right)\right]$$

$$\mu_2 = \frac{\lambda^2\alpha}{(\alpha-2)(\alpha-1)^2} \tag{6.1.6}$$

Thus, the **variance** of the **Pareto distribution** is $\frac{\lambda^2\alpha}{(\alpha-2)(\alpha-1)^2}; \alpha > 2$

The third central moment is

$$\mu_3 = \mu_3' - 3\mu_2'\mu_1' + 2\mu_1'^3$$

Similarly, we get the third central moment using the first three raw moments

$$= \frac{\lambda^3\alpha}{(\alpha-3)} - 3\frac{\lambda^2\alpha}{(\alpha-2)}\left[\frac{\lambda\alpha}{(\alpha-1)}\right] + 2\left[\frac{\lambda\alpha}{(\alpha-1)}\right]^3 \quad \text{[using Eqns. (6.1.1)–(6.1.3)]}$$

$$= \frac{\lambda^3\alpha}{(\alpha-3)(\alpha-2)(\alpha-1)^3}\left[(\alpha-2)(\alpha-1)^3 - 3\alpha(\alpha-3)(\alpha-1)^2 + 2\alpha^2(\alpha-3)(\alpha-2)\right]$$

$$= \frac{\lambda^3\alpha}{(\alpha-3)(\alpha-2)(\alpha-1)^3}\begin{bmatrix}\left(\alpha^4 - 3\alpha^3 - \alpha + 3\alpha^2 - 2\alpha^3 + 6\alpha^2 + 2 - 6\alpha\right) \\ -3\alpha\left(\alpha^3 - 2\alpha^2 + \alpha - 3\alpha^2 + 6\alpha - 3\right) + 2\left(\alpha^4 - 5\alpha^3 + 6\alpha^2\right)\end{bmatrix}$$

$$= \frac{\lambda^3\alpha}{(\alpha-3)(\alpha-2)(\alpha-1)^3}\begin{bmatrix}\left(\alpha^4 - 3\alpha^3 - \alpha + 3\alpha^2 - 2\alpha^3 + 6\alpha^2 + 2 - 6\alpha\right) \\ -3\alpha\left(\alpha^3 - 5\alpha^2 + 7\alpha - 3\right) + 2\left(\alpha^4 - 5\alpha^3 + 6\alpha^2\right)\end{bmatrix}$$

$$= \frac{\lambda^3\alpha}{(\alpha-3)(\alpha-2)(\alpha-1)^3}\begin{bmatrix}\alpha^4 - 5\alpha^3 + 9\alpha^2 - 7\alpha + 2 \\ -3\left(\alpha^4 - 5\alpha^3 + 7\alpha^2 - 3\alpha\right) + 2\left(\alpha^4 - 5\alpha^3 + 6\alpha^2\right)\end{bmatrix}$$

$$\mu_3 = \frac{2\lambda^3\alpha}{(\alpha-3)(\alpha-2)(\alpha-1)^3}(\alpha+1); \quad \alpha > 3 \tag{6.1.7}$$

Similarly, we get fourth central moment using the first four raw moments

$$\mu_4 = \mu_4' - 4\mu_3'\mu_1' + 6\mu_2'\mu_1'^2 - 3\mu_1'^4$$

$$= \frac{\lambda^4\alpha}{(\alpha-4)} - 4\frac{\lambda^3\alpha}{(\alpha-3)}\left[\frac{\lambda\alpha}{(\alpha-1)}\right] + 6\frac{\lambda^2\alpha}{(\alpha-2)}\left[\frac{\lambda\alpha}{(\alpha-1)}\right]^2 - 3\left[\frac{\lambda\alpha}{(\alpha-1)}\right]^4$$

[using Eqns. (6.1.1)–(6.1.4)]

$$= \frac{\lambda^4\alpha}{(\alpha-4)(\alpha-3)(\alpha-2)(\alpha-1)^4}$$

$$\begin{bmatrix}(\alpha-3)(\alpha-2)(\alpha-1)^4 - 4\alpha(\alpha-4)(\alpha-2)(\alpha-1)^3 \\ +6\alpha^2(\alpha-4)(\alpha-3)(\alpha-1)^2 - 3\alpha^3(\alpha-4)(\alpha-3)(\alpha-2)\end{bmatrix}$$

$$= \frac{\lambda^4\alpha}{(\alpha-4)(\alpha-3)(\alpha-2)(\alpha-1)^4}\begin{bmatrix}(\alpha-3)(\alpha-2)(\alpha^4+1-4\alpha^3+6\alpha^2-4\alpha)\\ -4\alpha(\alpha-4)(\alpha-2)(\alpha^3-1-3\alpha^2+3\alpha)\\ +6\alpha^2(\alpha-4)(\alpha-3)(\alpha^2-2\alpha+1)\\ -3\alpha^3(\alpha-4)(\alpha-3)(\alpha-2)\end{bmatrix}$$

$$= \frac{\lambda^4\alpha}{(\alpha-4)(\alpha-3)(\alpha-2)(\alpha-1)^4}\begin{bmatrix}(\alpha^2-5\alpha+6)(\alpha^4+1-4\alpha^3+6\alpha^2-4\alpha)\\ -4\alpha(\alpha^2-6\alpha+8)(\alpha^3-1-3\alpha^2+3\alpha)\\ +6\alpha^2(\alpha^2-7\alpha+12)(\alpha^2-2\alpha+1)\\ -3\alpha^3(\alpha^2-7\alpha+12)(\alpha-2)\end{bmatrix}$$

$$= \frac{\lambda^4\alpha}{(\alpha-4)(\alpha-3)(\alpha-2)(\alpha-1)^4}$$

$$\begin{bmatrix}\begin{pmatrix}\alpha^6+\alpha^2-4\alpha^5+6\alpha^4-4\alpha^3-5\alpha^5-5\alpha+20\alpha^4-30\alpha^3+20\alpha^2\\ +6\alpha^4+6-24\alpha^3+36\alpha^2-24\alpha\end{pmatrix}\\ -4\alpha\begin{pmatrix}\alpha^5-\alpha^2-3\alpha^4+3\alpha^3-6\alpha^4+6\alpha+18\alpha^3-18\alpha^2+8\alpha^3-8\\ -24\alpha^2+24\alpha\end{pmatrix}\\ +6\alpha^2(\alpha^4-2\alpha^3+\alpha^2-7\alpha^3+14\alpha^2-7\alpha+12\alpha^2-24\alpha+12)\\ -3\alpha^3(\alpha^3-7\alpha^2+12\alpha-2\alpha^2+14\alpha-24)\end{bmatrix}$$

$$= \frac{\lambda^4\alpha}{(\alpha-4)(\alpha-3)(\alpha-2)(\alpha-1)^4}\begin{bmatrix}(\alpha^6-9\alpha^5+32\alpha^4-58\alpha^3+57\alpha^2-29\alpha+6)\\ -4\alpha(\alpha^5-9\alpha^4+29\alpha^3-43\alpha^2+30\alpha-8)\\ +6\alpha^2(\alpha^4-9\alpha^3+27\alpha^2-31\alpha+12)\\ -3\alpha^3(\alpha^3-9\alpha^2+26\alpha-24)\end{bmatrix}$$

$$= \frac{\lambda^4\alpha}{(\alpha-4)(\alpha-3)(\alpha-2)(\alpha-1)^4}\begin{bmatrix}(\alpha^6-9\alpha^5+32\alpha^4-58\alpha^3+57\alpha^2-29\alpha+6)\\ -4(\alpha^6-9\alpha^5+29\alpha^4-43\alpha^3+30\alpha^2-8\alpha)\\ +6(\alpha^6-9\alpha^5+27\alpha^4-31\alpha^3+12\alpha^2)\\ -3(\alpha^6-9\alpha^5+26\alpha^4-24\alpha^3)\end{bmatrix}$$

$$\mu_4 = \frac{3\lambda^4\alpha}{(\alpha-4)(\alpha-3)(\alpha-2)(\alpha-1)^4}(3\alpha^2+\alpha+2); \quad \alpha > 4 \qquad (6.1.8)$$

c. **Skewness and Kurtosis of Pareto Distribution:** Using the Eqns. (6.1.6) and (6.1.7), we get

$$\beta_1 = \frac{\mu_3^2}{\mu_2^3}$$

$$= \frac{\left[\dfrac{2\lambda^3\alpha(\alpha+1)}{(\alpha-3)(\alpha-2)(\alpha-1)^3}\right]^2}{\left[\dfrac{\lambda^2\alpha}{(\alpha-2)(\alpha-1)^2}\right]^3}$$

$$\beta_1 = \frac{(\alpha-2)}{\alpha}\left(\frac{2(\alpha+1)}{(\alpha-3)}\right)^2$$

Using the above result β_1, we get

$$\gamma_1 = \frac{\mu_3}{\mu_2^{3/2}}$$

$$= \frac{\dfrac{2\lambda^3\alpha(\alpha+1)}{(\alpha-3)(\alpha-2)(\alpha-1)^3}}{\left[\dfrac{\lambda^2\alpha}{(\alpha-2)(\alpha-1)^2}\right]^{3/2}}$$

$$\gamma_1 = \sqrt{\frac{(\alpha-2)}{\alpha}}\left(\frac{2(\alpha+1)}{(\alpha-3)}\right)$$

Using the Eqns. (6.1.6) and (6.1.8), we get

$$\beta_2 = \frac{\mu_4}{\mu_2^2}$$

$$= \frac{\dfrac{3\lambda^4\alpha(3\alpha^2+\alpha+2)}{(\alpha-4)(\alpha-3)(\alpha-2)(\alpha-1)^4}}{\left[\dfrac{\lambda^2\alpha}{(\alpha-2)(\alpha-1)^2}\right]^2} \quad \text{[using Eqns. (6.1.6) and (6.1.8)]}$$

$$\beta_2 = \frac{3(\alpha-2)(3\alpha^2+\alpha+2)}{\alpha(\alpha-3)(\alpha-4)}$$

Using the above result β_2, we get

$$\gamma_2 = \beta_2 - 3$$

$$= \frac{3(\alpha-2)(3\alpha^2+\alpha+2)}{(\alpha-4)(\alpha-3)\alpha} - 3$$

$$= \frac{3}{(\alpha-4)(\alpha-3)\alpha}\left[(\alpha-2)(3\alpha^2+\alpha+2)-(\alpha-4)(\alpha-3)\alpha\right]$$

$$= \frac{3}{(\alpha-4)(\alpha-3)\alpha}\left[(3\alpha^3+\alpha^2+2\alpha-6\alpha^2-2\alpha-4)-(\alpha^2-7\alpha+12)\alpha\right]$$

$$= \frac{3}{(\alpha-4)(\alpha-3)\alpha}\left[(3\alpha^3-5\alpha^2-4)-(\alpha^3-7\alpha^2+12\alpha)\right]$$

$$\gamma_2 = \frac{6(\alpha^3-\alpha^2+6\alpha-2)}{\alpha(\alpha-3)(\alpha-4)}$$

d. **Median, Quartile, and Percentile of the Pareto Distribution:** Median divides the data into two equal parts, or in other words, we can write

$$F_X(M) = \frac{1}{2},$$

where M denotes the median.

From the above section, we know the CDF of Pareto distribution, so by substituting M in place of x in the expression of the CDF, we get

$$1-\left(\frac{\lambda}{M}\right)^{\alpha} = \frac{1}{2}$$

$$M = \lambda 2^{1/\alpha}$$

Quartiles divide the data into four equal parts, or in other words, we can write

$$F_X(Q_r) = \frac{r}{4} \qquad ; r = 1, 2, 3$$

where Q_r denotes the rth quantile.

$$Q_r = \frac{\lambda}{\left(1-\frac{r}{4}\right)^{1/\alpha}}$$

Clearly, Q_2 equals to M, i.e., the second quantile is equal to the median.

Percentiles divide the data into 100 equal parts, or in other words, we can write

$$F_X(P_r) = \frac{r}{100} \qquad r = 1, 2, \ldots, 99$$

where; P_r denotes the rth percentile.

$$P_r = \frac{\lambda}{\left(1 - \frac{r}{100}\right)^{1/\alpha}} \qquad r = 1, 2, \ldots, 99$$

We can find the expressions of P_1, P_2, …, P_{100} by putting r =1 to 99, respectively in the above expression.

e. Mode of the Pareto Distribution: Mode is the highest occurred value in the data set and to obtain the mode of this model, we have to check the behavior of its PDF. Now, consider the logarithmic of the PDF

$$\log f_X(x) = \log_e\left(\frac{\alpha\lambda^\alpha}{x^{\alpha+1}}\right), \qquad x > \lambda$$

$$\log f_X(x) = \log_e \alpha + \alpha \log_e \lambda - (\alpha+1)\log_e x, \qquad x > \lambda$$

Differentiating both the side with respect to x

$$\frac{1}{f_X(x)} f_X'(x) = -\frac{(\alpha+1)}{x}$$

$$f_X'(x) = -\left[\frac{(\alpha+1)}{x}\right] f_X(x) < 0, \qquad \forall x > \lambda$$

This implies that $f(x)$ decreasing function of x. Therefore, mode = λ

f. Reliability Function of Pareto Distribution

$$R(x) = P[X > x] = 1 - F_X(x)$$

$$R(x) = \begin{cases} \left(\frac{\lambda}{x}\right)^\alpha & ; x > \lambda \\ 1 & ; x \le \lambda \end{cases}$$

g. Hazard Rate of Pareto Distribution

The hazard (or failure) rate $\lambda(x)$ of this model is given by

$$\lambda(x) = \frac{1}{R(x)} f(x)$$

Substituting the expression of $R(x)$ and $f(x)$, we get

$$\lambda(x) = \frac{1}{\left(\frac{\lambda}{x}\right)^{\alpha}} \alpha \lambda^{\alpha} \left(\frac{1}{x}\right)^{\alpha+1}$$

$$\lambda(x) = \frac{\alpha}{x} \quad ; x > \lambda$$

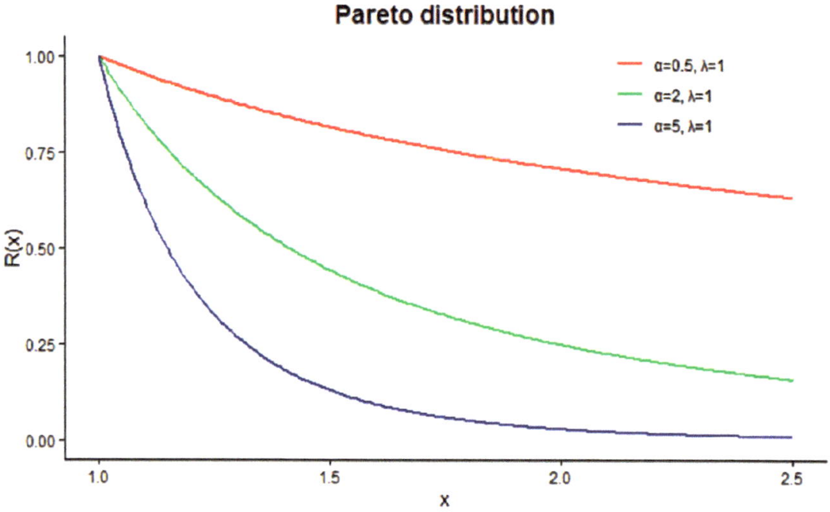

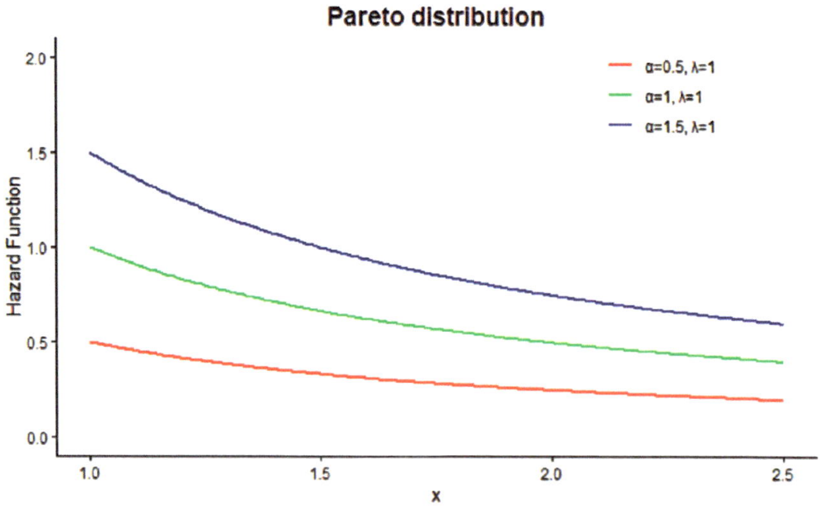

FIGURE 6.2 Reliability function and hazard function of Pareto distribution.

h. Mean Time to Failure of Pareto Distribution: In reliability terminology, the mean of the random variable X in the absence of repair and replacement is known as the meantime to failure (MTTF). The expression of MTTF is the same as the mean of the model, i.e., $\text{MTTF} = \frac{\alpha\lambda}{(\alpha-1)}, \alpha > 1.$

6.2 GENERALIZED PARETO DISTRIBUTION

If a random variable X follows Pareto distribution, then the expression of the CDF is

$$F_X(x) = 1 - \left(\frac{\lambda}{\lambda + x}\right)^{\alpha}; \qquad 0 < x < \infty$$

Symbolically: $X \sim P(\alpha, \lambda)$

Here; α, λ (>0) are the parameters of the distribution.

The density function is

$$f_X(x) = \frac{d}{dx} F_X(x) = \frac{d}{dx}\left\{1 - \left(\frac{\lambda}{\lambda + x}\right)^{\alpha}\right\}$$

$$f_X(x) = \frac{\alpha\lambda^{\alpha}}{(\lambda + x)^{\alpha+1}}\ ;\ x > 0,\ \alpha > 0,\ \lambda > 0.$$

6.2.1 PROPERTIES OF THE GENERALIZED PARETO DISTRIBUTION

a. Moments of Generalized Pareto Distribution

The first four raw moments $\mu_1', \mu_2', \mu_3', \mu_4'$ have been derived below

$$\mu_1' = E(X)$$

$$= \int_0^{\infty} x\alpha \frac{\lambda^{\alpha}}{(\lambda + x)^{\alpha+1}} dx$$

$$= \alpha\lambda^{\alpha} \int_0^{\infty} x \frac{1}{(\lambda + x)^{\alpha+1}} dx$$

$$Let\ \ \lambda + x = z \Rightarrow dx = dz\ \&\ x = z - \lambda$$

Substituting the values of x and dx, we get

$$= \alpha\lambda^{\alpha}\int_{\lambda}^{\infty}(z-\lambda)\frac{1}{z^{\alpha+1}}\,dz$$

$$= \alpha\lambda^{\alpha}\int_{\lambda}^{\infty}z^{-\alpha}\,dz - \alpha\lambda^{\alpha+1}\int_{\lambda}^{\infty}z^{-\alpha-1}\,dz$$

$$= \alpha\lambda^{\alpha}\left[\frac{z^{-\alpha+1}}{(-\alpha+1)}\right]_{\lambda}^{\infty} - \alpha\lambda^{\alpha+1}\left[\frac{z^{-\alpha}}{(-\alpha)}\right]_{\lambda}^{\infty}$$

$$= \alpha\lambda^{\alpha}\left[\frac{\lambda^{-\alpha+1}}{(-\alpha+1)}\right]_{\lambda}^{\infty} - \alpha\lambda^{\alpha+1}\left[\frac{\lambda^{-\alpha}}{(-\alpha)}\right]_{\lambda}^{\infty}$$

$$= \frac{\alpha\lambda}{(\alpha-1)} + \lambda$$

$$\mu_1' = \frac{\lambda}{(\alpha-1)}; \alpha > 1$$

Thus, the *mean* of the Generalized Pareto distribution is $\frac{\lambda}{(\alpha-1)}$.

Second raw moment

$$\mu_2' = E(X^2)$$

$$= \int_{0}^{\infty}x^2\alpha\frac{\lambda^{\alpha}}{(\lambda+x)^{\alpha+1}}\,dx$$

$$= \alpha\lambda^{\alpha}\int_{0}^{\infty}x^2\frac{1}{(\lambda+x)^{\alpha+1}}\,dx$$

$$Let\ \lambda + x = z \Rightarrow dx = dz\ \&\ x = z - \lambda$$

Substituting the values of x and dx, we get

$$= \alpha\lambda^{\alpha}\int_{\lambda}^{\infty}(z-\lambda)^2\frac{1}{z^{\alpha+1}}\,dz = \alpha\lambda^{\alpha}\int_{\lambda}^{\infty}(z^2+\lambda^2-2z\lambda)\frac{1}{z^{\alpha+1}}\,dz$$

$$= \alpha\lambda^{\alpha}\int_{\lambda}^{\infty}z^{-\alpha+1}\,dz + \alpha\lambda^{\alpha+2}\int_{\lambda}^{\infty}z^{-\alpha-1}\,dz - 2\alpha\lambda^{\alpha+1}\int_{\lambda}^{\infty}z^{-\alpha}\,dz$$

$$= \alpha\lambda^{\alpha}\left[\frac{z^{-\alpha+2}}{(-\alpha+2)}\right]_{\lambda}^{\infty} + \alpha\lambda^{\alpha+2}\left[\frac{z^{-\alpha}}{(-\alpha)}\right]_{\lambda}^{\infty} - 2\alpha\lambda^{\alpha+1}\left[\frac{z^{-\alpha+1}}{(-\alpha+1)}\right]_{\lambda}^{\infty}$$

$$= \alpha\lambda^{\alpha}\left[\frac{\lambda^{-\alpha+2}}{(\alpha-2)}\right] + \alpha\lambda^{\alpha+2}\left[\frac{\lambda^{-\alpha}}{(\alpha)}\right] - 2\alpha\lambda^{\alpha+1}\left[\frac{\lambda^{-\alpha+1}}{(\alpha-1)}\right]$$

$$= \alpha\left[\frac{\lambda^{2}}{(\alpha-2)}\right] + \alpha\left[\frac{\lambda^{2}}{(\alpha)}\right] - 2\alpha\left[\frac{\lambda^{2}}{(\alpha-1)}\right]$$

$$= \alpha\lambda^{2}\left\{\frac{1}{(\alpha-2)} - \frac{1}{\alpha} - 2\frac{1}{1-\alpha}\right\}$$

$$= \frac{\lambda^{2}}{(\alpha-1)(\alpha-2)}\left(\alpha - \alpha^{2} - \alpha^{2} + 3\alpha - 2 - 4\alpha + 2\alpha^{2}\right)$$

$$\mu_2' = \frac{2\lambda^{2}}{(\alpha-1)(\alpha-2)};\alpha > 2$$

c. **Reliability Function, Hazard Rate, Mean Time to Failure of Generalized Pareto Distribution**

$$R(x) = \begin{cases} \left(\dfrac{\lambda}{x+\lambda}\right)^{\alpha} & ;x > 0 \\ 1 & ;x \le 0 \end{cases}$$

$$\lambda(x) = \alpha\frac{1}{(x+\lambda)} \quad ;x > 0$$

$$\text{MTTF} = \frac{\lambda\alpha}{(\alpha-1)} \quad ;\alpha > 1$$

6.3 BURR DISTRIBUTION

If a random variable X follows Burr distribution, then the expression of the CDF is:

$$F_X(x) = 1 - \left(\frac{1}{1+x^{\gamma}}\right)^{\alpha}; \qquad 0 < x < \infty$$

Here α, γ (>0) are the parameters of the distribution.

The density function is

$$f_X(x) = \frac{d}{dx}F_X(x) = \frac{d}{dx}\left\{1 - \left(\frac{1}{1+x^{\gamma}}\right)^{\alpha}\right\}$$

$$f_X(x) = \frac{\alpha\gamma x^{\gamma-1}}{(1+x^{\gamma})^{\alpha+1}} \; ; x > 0, \; \alpha > 0, \gamma > 0$$

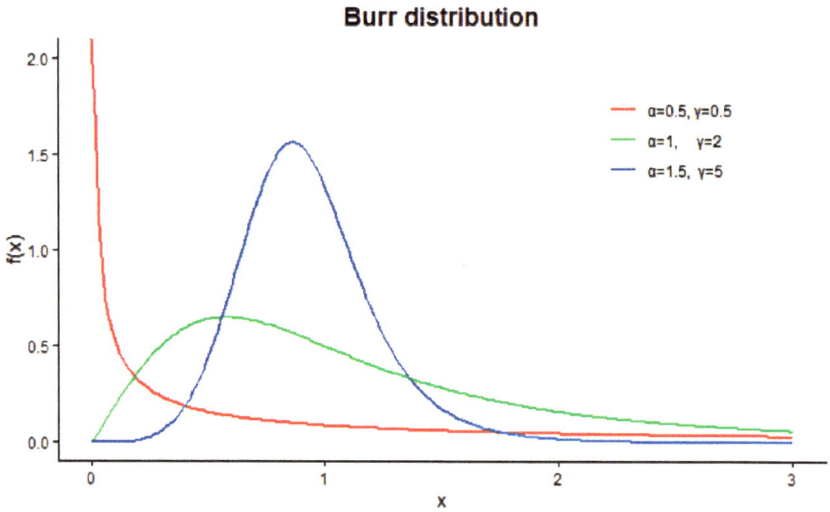

FIGURE 6.3 Burr distribution.

Sample Generation of Using Inverse Transformation Method:

$$F_X(x) = 1 - \left(\frac{1}{1+x^{\gamma}}\right)^{\alpha} \; ; \qquad x > 0$$

$$U = 1 - \left(\frac{1}{1+x^{\gamma}}\right)^{\alpha}$$

$$1 - U = \left(\frac{1}{1+x^{\gamma}}\right)^{\alpha}$$

$$x = \left[(1-U)^{-1/\alpha} - 1\right]^{1/\gamma}$$

where U follows uniform (0,1).

➢ **Note:** Generate a random sample of size n from $U(0,1)$ i.e., $\underline{u} = (u_1, u_2, \ldots, u_n)$ (say). Define, $x_i = \left[(1-u_i)^{-1/\alpha} - 1\right]^{1/\gamma}$; $i = 1, 2, .., n.$ Then, $\underline{x} = (x_1, x_2, \ldots, x_n)$ gives a random sample of size n from Burr(α, γ).

- **Example 6.2:** Let u = (0.1,0.3,0.22,0.51,0.11,0.6) be a random sample from $U(0,1)$ distribution. Now, we want to generate a random sample from Burr distribution. Calculate $x_i = \left[(1-u_i)^{-1/\alpha} - 1\right]^{1/\gamma}$; $i = 1,2,..,6$, where u_i denotes the ith element of $(u_1,u_2,\ldots,u_n)$.

***U*(0,1)**	Random Sample		
	***Burr*(2,3)**	***Burr*(2,5)**	***Burr*(3,5)**
0.10	0.3781921	0.5579914	0.5136190
0.30	0.5801155	0.7212880	0.6610660
0.22	0.5095203	0.6672628	0.6127018
0.51	0.7539474	0.8441209	0.7687190
0.11	0.3914822	0.5696750	0.5242743
0.60	0.8345006	0.8971312	0.8139251

6.3.1 PROPERTIES OF THE BURR DISTRIBUTION

r^{th} raw moments:

$$\mu_r' = E\left(X^r\right)$$

$$= \int_0^\infty x^r \frac{\alpha\gamma x^{\gamma-1}}{(1+x^\gamma)^{\alpha+1}} dx$$

$$= \alpha\gamma \int_0^\infty \frac{x^{\gamma+r-1}}{(1+x^\gamma)^{\alpha+1}} dx$$

$$Let \;\; x^\gamma = z \Rightarrow x = (z)^{1/\gamma} \;\; \& \; dx = 1/\gamma (z)^{(1/\gamma)-1} dz$$

$$= \alpha\gamma \int_0^\infty \frac{z[z]^{(r-1)/\gamma}}{(1+z)^{\alpha+1}} 1/\gamma (z)^{(1/\gamma)-1} dz$$

$$= \alpha \int_0^\infty \frac{1}{(1+z)^{\alpha+1}} (z)^{(r/\gamma)} dz$$

Now, using the results of the Beta function of the second kind, we get

$$\mu_r' = \alpha B\left(\alpha - \frac{r}{\gamma}, \frac{r}{\gamma} + 1\right), \quad \alpha > \frac{r}{\gamma}$$

For $r = 1$, $\mu_1' = \alpha B\left(\alpha - \frac{1}{\gamma}, \frac{1}{\gamma} + 1\right), \quad \alpha > \frac{1}{\gamma}.$

Therefore, the mean of the Burr distribution is

$$\alpha B\left(\alpha - \frac{1}{\gamma}, \frac{1}{\gamma} + 1\right)$$

For $r = 2$ $\mu_2' = \alpha B\left(\alpha - \frac{2}{\gamma}, \frac{2}{\gamma} + 1\right), \quad \alpha > \frac{2}{\gamma}.$

The **variance** of the Burr distribution is given below

$$\mu_2 = \mu_2' - \mu_1'^2$$

$$\mu_2 = \alpha B\left(\alpha - \frac{2}{\gamma}, \frac{2}{\gamma} + 1\right) - \left[\alpha B\left(\alpha - \frac{1}{\gamma}, \frac{1}{\gamma} + 1\right)\right]^2$$

b. **Median, Quartile, and Percentile of the Burr Distribution:** Median divides the data into two equal parts, or in other words, we can write

$$F_X(M) = \frac{1}{2}$$

where M denotes the median.

From the above section, we know the CDF of Burr distribution, so by substituting M in place of x in the expression of the CDF, we get

$$1 - \left(\frac{1}{1 + M^\gamma}\right)^\alpha = \frac{1}{2}$$

$$\Rightarrow \left(\frac{1}{1 + M^\gamma}\right) = \left(\frac{1}{2}\right)^{1/\alpha}$$

$$\Rightarrow 1 + M^\gamma = 2^{1/\alpha}$$

$$\Rightarrow M = \left[\left(2^{1/\alpha} - 1\right)\right]^{1/\gamma}$$

Quartiles divide the data into four equal parts, or in other words, we can write

$$F_X(Q_r) = \frac{r}{4} \qquad r = 1, 2, 3$$

where Q_r denotes the rth quantile.

$$Q_r = \left[\frac{1}{\left(1 - \frac{r}{4}\right)^{1/\alpha}} - 1 \right]^{1/\gamma}$$

Clearly, Q_2 equals to M, i.e., the second quantile is equal to the median.

Percentiles divide the data into 100 equal parts, or in other words, we can write

$$F_X(P_r) = \frac{r}{100} \qquad r = 1, 2, \ldots, 99$$

where P_r denotes the rth percentile.

$$P_r = \left[\frac{1}{\left(1 - \frac{r}{100}\right)^{1/\alpha}} - 1 \right]^{1/\gamma} \qquad r = 1, 2, \ldots, 99$$

We can find the expressions of P_1, P_2, …, P_{100} by putting r =1 to 99, respectively in the above expression.

c. **Mode of the Burr Distribution:** Mode is the highest occurred value in the data set and to obtain the mode of this model, we have differentiated the PDF with respect to x, which is given below

$$\log f_X(x) = \log_e \left(\frac{\alpha \gamma x^{\gamma-1}}{(1+x^\gamma)^{\alpha+1}} \right)$$

$$\log f_X(x) = \log_e \alpha + \log_e \gamma + (\gamma - 1)\log_e x - (\alpha + 1)\log_e (1 + x^\gamma)$$

Differentiating both the side with respect to x

$$\frac{1}{f_X(x)} f_X'(x) = \frac{(\gamma - 1)}{x} - \frac{(\alpha + 1)}{(1 + x^\gamma)} \gamma x^{\gamma-1}$$

$$f_X'(x) = \left[\frac{(\gamma - 1)}{x} - \frac{(\alpha + 1)}{(1 + x^\gamma)} \gamma x^{\gamma-1} \right] f_X(x)$$

Since $f'_X(x) = 0$ if

$$\left[\frac{(\gamma-1)}{x} - \frac{(\alpha+1)}{(1+x^{\gamma})}\gamma x^{\gamma-1}\right] f_X(x) = 0$$

The density function always positive.

$$\Rightarrow \frac{(\gamma-1)}{x} - \frac{(\alpha+1)}{(1+x^{\gamma})}\gamma x^{\gamma-1} = 0$$

$$x^{\gamma} = \frac{-(\gamma-1)}{((\gamma-1)-(\alpha+1)\gamma)}$$

$$\text{mode} = \left(\frac{\gamma-1}{1+\alpha\gamma}\right)^{1/\gamma}$$

d. **Reliability Function of Burr Distribution:** The reliability function $R(x)$ at the time (x) of the component/system is given by

$R(x)$ = Probability that the time to failure of the component/system is greater than x

$$R(x) = P[X > x] = 1 - F_X(x)$$

$$R(x) = \left(\frac{1}{1+x^{\gamma}}\right)^{\alpha}, \qquad 0 < x < \infty$$

f. **Hazard Rate of Burr Distribution:** The hazard (or failure) rate of this model is given by

$$\lambda(x) = \frac{1}{R(x)} f(x)$$

Substituting the expression of $R(x)$ and $f(x)$, we get:

$$\lambda(x) = \frac{1}{\left(\frac{1}{1+x^{\gamma}}\right)^{\alpha}} \frac{\alpha\gamma x^{\gamma-1}}{(1+x^{\gamma})^{\alpha+1}}$$

$$\lambda(x) = \frac{\alpha\gamma x^{\gamma-1}}{(1+x^{\gamma})} \; ; x > 0$$

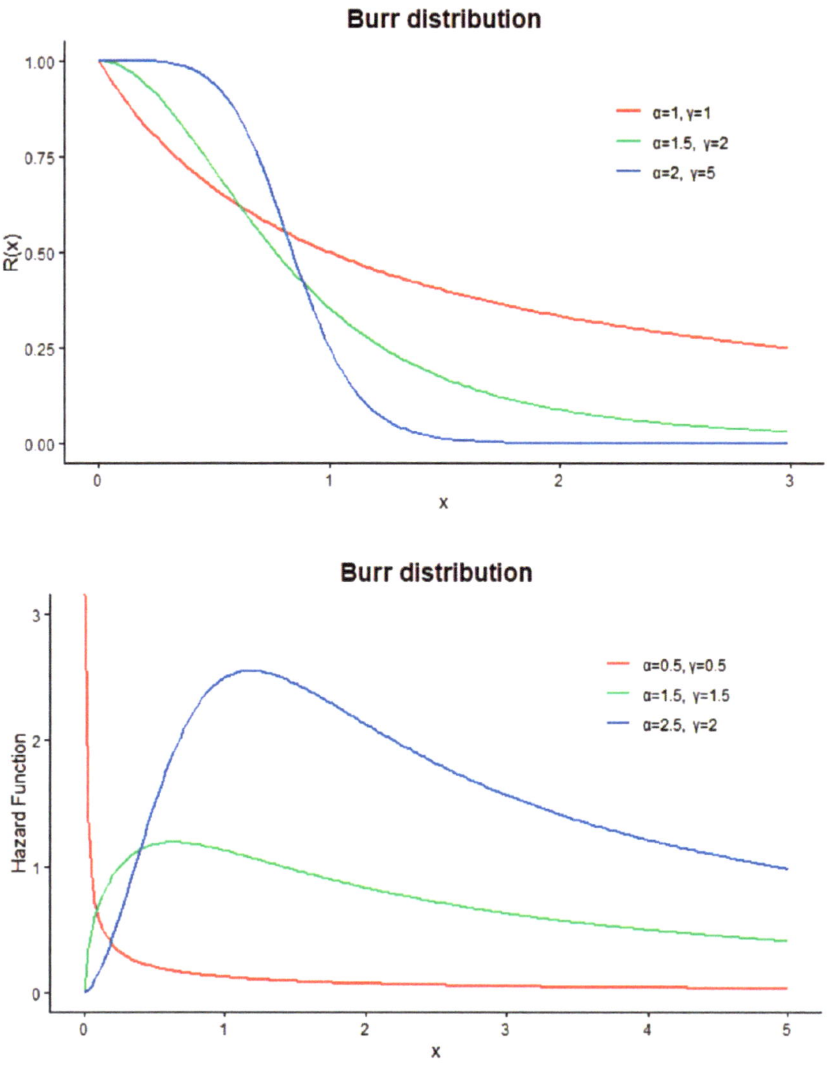

FIGURE 6.4 Reliability function and hazard function of Burr distribution.

g. **Mean Time to Failure of Burr Distribution:** The expression of MTTF of this model is

$$\text{MTTF} = \alpha B\left(\alpha - \frac{1}{\gamma}, \frac{1}{\gamma} + 1\right), \quad \alpha > \frac{1}{\gamma}$$

KEYWORDS

- **Cumulative distribution function**
- **Meantime to failure**
- **Pareto distribution**
- **Probability density function**

REFERENCES

Alzaatreh, A., Famoye, F., & Lee, C., (2013). Weibull-pareto distribution and its applications. *Communication in Statistics- Theory and Methods, 42*(42), 1673–1691.

Hakim, A., Fithriani, I., & Novita, M., (2021). Properties of Burr distribution and its application to heavy-tailed survival time data. *Journal of Physics: Conference Series, 1725*. 012016. 10.1088/1742-6596/1725/1/012016.

Ibrahim, E., & Gokarna, A., (2017). A new generalization of the exponential Pareto distribution. *Journal of Information and Optimization Sciences, 38*(5), 675–697.

Jayakumar, K., Bindu, K., & Hamedani, G. G., (2020). On a new generalization of Pareto distribution and its applications. *Communications in Statistics-Simulation and Computation, 49(*5), 1264–1284.

Moharram, S. H., Gosain, A. K., & Kapoor, P. N., (1993). A comparative study for the estimators of the generalized Pareto distribution. *Journal of Hydrology, 150*(1), 169–185.

CHAPTER 7

Dynamics of Data Analysis and Visualization with Python

PRASHANT VERMA[1], SHWETA DIXIT[2], MUKTI KHETAN[3], SURESH BADARLA[3], and FOZIA HOMA[4]

[1]Department of Statistics, Faculty of Science, University of Allahabad, Prayagraj, India

[2]Clinical Development Services Agency, Translational Health Science and Technology Institute, NCR Biotech Science Cluster, Haryana, India

[3]Department of Mathematics, Amity School of Applied Sciences, Amity University Mumbai, Maharashtra, India

[4]Department of Statistics, Mathematics, and Computer Application, Bihar Agricultural University, Sabour, Bhagalpur, Bihar, India

7.1 BACKGROUND

In 1989, Guido Van Rossum developed an object-oriented programming language – Python. It has been optimally designed for rapid prototyping of complex applications in the field of data science. It has interfaces to several Operating Systems and libraries. Considering the large-scale utilities of Python, many worldwide renowned companies and organizations use the

Distribution Theory: Principles and Applications. Fozia Homa, Mukti Khetan, Mohd. Arshad, & Pradeep Mishra (Naz) (Eds.)

Python programming language, e.g., GlaxoSmithKline (GSK), ISRO, NASA, Google, YouTube, and Bit-Torrent.

Python programming is widely used in Machine Learning, Natural Language Processing, Artificial Intelligence, Neural Networks, and other advanced dimensions of Information Technology and Computer Science. Code readability has been the primary center of attention for the development of Python. This chapter will demonstrate the applications of Python in reading data, data visualization, data manipulation, data analysis, model building and prediction using an open-source data set.

7.2 INSTALLING PYTHON

Python is open source and is free to install. Before we start working with Python, we need first to download the installation package of the desired version of Python from the link: https://www.python.org/downloads/. This page will direct the user to choose between the two latest versions for Python 2 and 3: Python 3.9.1 and Python 2.7.18. Alternatively, if a user needs a specific Python version, the page can be scrolled down to find download links for earlier versions.

The current exercise will be done in 'Jupyter Notebook;' however, one can use 'Spyder' or other programming interfaces using the same programming codes for Python.

7.3 INSTALLING PYTHON LIBRARIES

We can use Anaconda prompt for installing a new library using the code "pip install library_name." Alternatively, the libraries can be installed through Jupyter Notebook as well. Once the library is installed, the same needs to be imported using the import command – "import library_name as preffered_name." Usually, the preferred name of the library is taken as the initials or short names of the library, e.g., "import numpy as np."

7.3.1 IMPORTING PANDAS LIBRARY

In the Python programming language, pandas refer to a software library developed for data analysis and manipulation. For example, pandas propound data structures and operations for manipulating Data frames or numerical tables.

```
import pandas as pd
```

7.4 IMPORT DATA FROM .CSV FILE

The Titanic data set is a renowned piece of data that is open source. The data set contains the information of passengers onboard the Titanic, a British passenger liner that sank in the North Atlantic Ocean in 1912. This data set is often used to demonstrate the binary classification problem in which we try to predict if a passenger boarded on the ship survives or not. The data is available in .csv format. Thus the data will be fetched using the below codes.

```
Data = pd.read_csv("C:\\Users\\Desktop\\Desktop\\Py\\Titan_Data\\Data_Titanic.csv")
```

In Python, we can choose single quotes (' ') as well as double quotes (" ") to address the location of the file name, variable name, and other functions. A user can read the excel file using '_excel' command in place of '_csv' command.

7.4.1 RENAMING THE DATASET

Suppose we need to change the name of our dataset from 'Data' to 'df,' the below command will serve the purpose.

```
df=Data
```

The dataset 'Data' now be stored as a different data frame 'df.' In fact, Python will create a duplicate data frame with the name 'df.' Thus, we can use both 'df' and 'data' data-frames for further analysis.

For data preview, we execute the below code.

```
df.head()
```

	PassengerId	Survived	Pclass	Name	Sex	Age	SibSp	Parch	Ticket	Fare	Cabin	Embarked
0	1	0	3	Braund, Mr. Owen Harris	male	22.0	1	0	A/5 21171	7.2500	NaN	S
1	2	1	1	Cumings, Mrs. John Bradley (Florence Briggs Th	female	38.0	1	0	PC 17599	71.2833	C85	C
2	3	1	3	Heikkinen, Miss. Laina	female	26.0	0	0	STON/O2. 3101282	7.9250	NaN	S
3	4	1	1	Futrelle, Mrs. Jacques Heath (Lily May Peel)	female	35.0	1	0	113803	53.1000	C123	S
4	5	0	3	Allen, Mr. William Henry	male	35.0	0	0	373450	8.0500	NaN	S

If we wish to preview the desired number of observations (row), e.g., 10 rows of the data frame (df), we can execute the below code.

```
df.head(10)
```

PassengerId	Survived	Pclass	Name	Sex	Age	SibSp	Parch	Ticket	Fare	Cabin	Embarked
1	0	3	Braund, Mr. Owen Harris	male	22.0	1	0	A/5 21171	7.2500	NaN	S
2	1	1	Cumings, Mrs. John Bradley (Florence Briggs Th...	female	38.0	1	0	PC 17599	71.2833	C85	C
3	1	3	Heikkinen, Miss. Laina	female	26.0	0	0	STON/O2. 3101282	7.9250	NaN	S
4	1	1	Futrelle, Mrs. Jacques Heath (Lily May Peel)	female	35.0	1	0	113803	53.1000	C123	S
5	0	3	Allen, Mr. William Henry	male	35.0	0	0	373450	8.0500	NaN	S
6	0	3	Moran, Mr. James	male	NaN	0	0	330877	8.4583	NaN	Q
7	0	1	McCarthy, Mr. Timothy J	male	54.0	0	0	17463	51.8625	E46	S
8	0	3	Palsson, Master. Gosta Leonard	male	2.0	3	1	349909	21.0750	NaN	S
9	1	3	Johnson, Mrs. Oscar W (Elisabeth Vilhelmina Berg)	female	27.0	0	2	347742	11.1333	NaN	S
10	1	2	Nasser, Mrs. Nicholas (Adele Achem)	female	14.0	1	0	237736	30.0708	NaN	C

7.4.2 SELECTING A RANDOM SAMPLE WITH REPLACEMENT

A random sample with replacement is taken using the codes mentioned below. This new data frame Data_Titanic consists of 95% observations of the original dataset **df**.

```
Data_Titanic=df.sample(frac=0.95, replace=True, random_state=1)
```

Now further, we will be dealing with this dataset 'Data_Titanic.'

7.5 DATA EXPLORATION AND PREPARATION

In data analysis, data exploration is the initial step and basically involves summarizing the foremost attributes of a data set, including its shape, accuracy, initial patterns in the dataset, and other features.

Followings are the steps involved to understand, clean, and prepare the data for analysis and model building:

i. Variable identification;
ii. Univariate analysis;
iii. Bi-variate analysis;
iv. Missing values treatment;
v. Variable transformation;
vi. Variable creation.

The data dictionary is the most significant tool to understand the data for further treatments. The data dictionary is also called the metadata, a set of data that narrate and furnish information about other data which we are going to use for analysis or model building. To understand the below data dictionary let us have a preview of the data Data_Titanic.

```
Data_Titanic.head(5)
```

PassengerId	Survived	Pclass	Name	Sex	Age	SibSp	Parch	Ticket	Fare	Cabin	Embarked
38	0	3	Cann, Mr. Ernest Charles	male	21.0	0	0	A./5. 2152	8.05	NaN	S
236	0	3	Harknett, Miss. Alice Phoebe	female	NaN	0	0	W./C. 6609	7.55	NaN	S
73	0	2	Hood, Mr. Ambrose Jr	male	21.0	0	0	S.O.C. 14879	73.50	NaN	S
768	0	3	Mangan, Miss. Mary	female	30.5	0	0	364850	7.75	NaN	Q
716	0	3	Soholt, Mr. Peter Andreas Lauritz Andersen	male	19.0	0	0	348124	7.65	F G73	S

➢ **Data Dictionary Details:**

1. Survived: 0 = No, 1 = Yes (Survival of the Passengers)
2. pclass: Economic status
   1st = Upper 2nd = Middle 3rd = Lower
3. Sex: Sex of the Passenger
4. Age: Age in years
5. sibsp: Number of siblings / spouses aboard the Titanic
6. parch: Number of parents / children aboard the Titanic
7. ticket: Ticket number
8. fare: Cost of the Ticket
9. cabin: Cabin number
10. embarked: Port of Embarkation; C = Cherbourg, Q = Queenstown, S = Southampton

7.5.1 DATA QUALITY AND MISSING VALUE ASSESSMENT

We always need an exemplary data set to perform data analysis and modeling; thus, we need to measure data quality effectively to serve the purpose. Data quality refers to the capability of a data set to achieve the desired objective. Effective data analysis cannot be done with low-quality data. There are several advanced strategies that we can use to improve the quality of the data and make the best use of it to fulfill our analytical requirements.

Before going forward to access the data quality, we should check the dimension of our dataset. The below code exhibits the dimensions of the given Data Frame. Here, it shows that the data set contains 846 passengers (rows) and 12 variables (columns).

```
Data_Titanic.shape
```

➢ **Output:** (846, 12)

Further, we can use the "describe" function to obtain the basic statistics of the data frame.

```
Data_Titanic.describe()
```

	PassengerId	Survived	Pclass	Age	SibSp	Parch	Fare
count	846.000000	846.000000	846.000000	675.000000	846.000000	846.000000	846.000000
mean	450.786052	0.338061	2.328605	29.809644	0.523641	0.386525	30.097168
std	255.867871	0.473329	0.818895	14.638398	1.116327	0.789432	41.069624
min	1.000000	0.000000	1.000000	0.420000	0.000000	0.000000	0.000000
25%	231.000000	0.000000	2.000000	20.750000	0.000000	0.000000	7.895800
50%	453.500000	0.000000	3.000000	28.000000	0.000000	0.000000	13.500000
75%	669.750000	1.000000	3.000000	39.000000	1.000000	0.000000	30.000000
max	891.000000	1.000000	3.000000	80.000000	8.000000	6.000000	263.000000

If we desire to get the basic statistics of a particular variable, then the below code can be executed using the describe function and the variable name. In the below command, basic statistics for the variable 'Age' have been obtained.

```
Data_Titanic['Age'].describe()
```

```
count    675.000000
mean      29.809644
std       14.638398
min        0.420000
25%       20.750000
50%       28.000000
75%       39.000000
max       80.000000
Name: Age, dtype: float64
```

In addition to getting all the basic statistics of a variable, one can find the value of a particular characteristic (parameter) corresponding to the variable. The below command enables us to obtain the minimum of the variable 'Age.'

```
Data_Titanic['Age'].min()
```

➤ **Output: 0.42**

The above value of age is obtained in years. Since it is in years, we can convert the age into months using the below code.

```
Data_Titanic['Age'].min()*12
```

➤ **Output: 5.04**

In order to realize the missing value and have a preview of missing values in the data frame, isnull() function can be used as below.

```
Data_Titanic.isnull().head()
```

	PassengerId	Survived	Pclass	Name	Sex	Age	SibSp	Parch	Ticket	Fare	Cabin	Embarked
37	False	False	False	False	False	False	False	False	False	False	True	False
235	False	False	False	False	False	True	False	False	False	False	True	False
72	False	False	False	False	False	False	False	False	False	False	True	False
767	False	False	False	False	False	False	False	False	False	False	True	False
715	False	False	False	False	False	False	False	False	False	False	False	False

In the above output, the "True" value reflects that the variable has the null value at that place. On the other hand, if we need to realize the count of missing observations corresponding to the various variables, the below code can serve the purpose.

```
Data_Titanic.isnull().sum()
```

```
PassengerId      0
Survived         0
Pclass           0
Name             0
Sex              0
Age            171
SibSp            0
Parch            0
Ticket           0
Fare             0
Cabin          665
Embarked         1
dtype: int64
```

7.5.1.1 DATA VISUALIZATION

Data visualization provides accessibility to find and understand patterns, outliers, and trends in the dataset. Charts, graphs, and maps are the most common data visualization tools.

Let us see how the 'Age' variable looks like in the Histogram (Figure 7.1) using **matplotlib.pyplot** library.

```
import matplotlib.pyplot as plt
plt.rc('font', size=11)
```

We can change the current font size (11) in the syntax as per the visualization need.

```
age_hist = Data_Titanic['Age'].hist(bins=15, color='cadetblue', alpha=.9)
age_hist.set(xlabel='Age', ylabel='Count')

plt.savefig('Hist_Age.jpg')
plt.show()
```

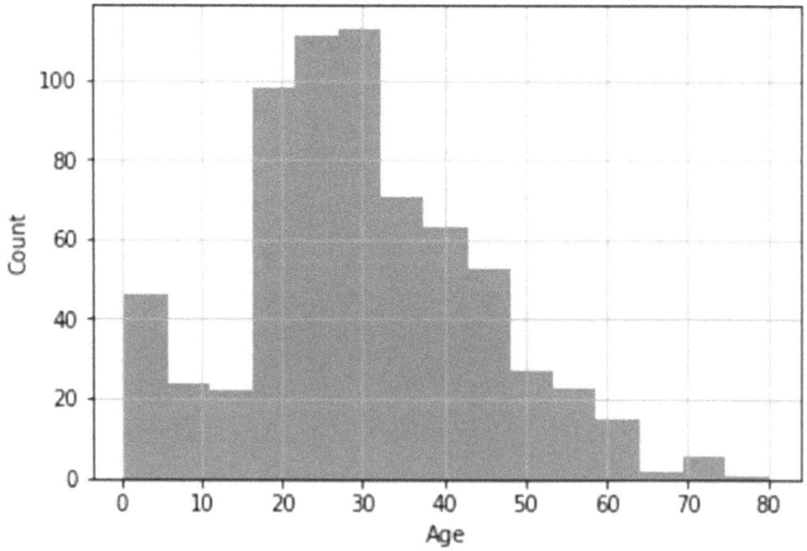

FIGURE 7.1 Histogram for the age distribution.

Referring to the above codes for a histogram, the following points are suggested:

i. Bins can be modified to show the width of the Histogram, color can be modified as per requirement, and alpha shows the thickness of the color (1 stand for 100% Thickness).
ii. We can choose a different file type, e.g., pdf or png, for exporting the image. Also, we can opt for the save image as an option by using right-click on the graph.
iii. Please note that the syntax 'plt.show()' should be placed after the 'plt.save' command; otherwise, the image may be saved as blank in the output file.

Furthermore, we have removed the grid lines in Histogram (Figure 7.2) using the below code:

```
age_hist = Data_Titanic['Age'].hist(bins=15, color='cadetblue', alpha=.9)
age_hist.set(xlabel='Age', ylabel='Count')
age_hist.grid(False)
plt.savefig('Hist_Age.jpg')
plt.show()
```

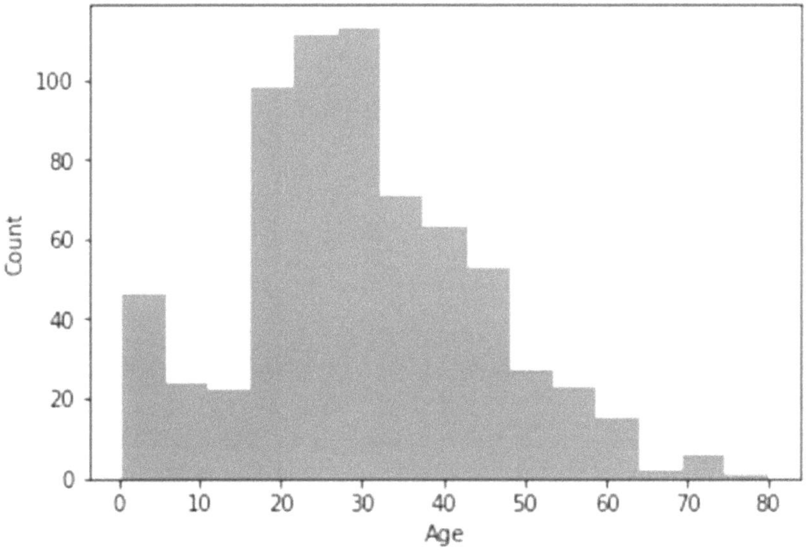

FIGURE 7.2 Histogram (without grid) for the age distribution.

➢ **Age–Missing Values:** Moving forward, we can find the missing value count corresponding to a particular variable. The below code supplies the count of null values found in 'Age' Variable.

```
Data_Titanic['Age'].isnull().sum()
```

➢ **Output: 171**

The above output suggests that in the data set total of 171 records are there with missing age values. In the below syntax, we have calculated the missing data proportion in 'Age' variable. First, we have calculated the total number of observations (846) in the data set. Then divide the number of missing records (171) by the total number of passengers.

```
len(Data_Titanic)
```

➢ **Output: 846**

```
171/len(Data_Titanic)
```

➢ **Output: 0.20212765957446807**

One may give the round-off value using the 'round' function as below. The proportion of missing values has been rounded off to the 3 decimal place.

```
round(171/len(Data_Titanic),3)
```

➢ **Output: 0.202**

Here, we can comment that ~20% of records do not have the age information of the passenger.

```
round(Data_Titanic['Age'].mean(),2)
```

➢ **Output: 29.81**

After glancing at the Histogram of 'Age,' we can observe that this variable is significantly skewed. This indicates that the mean might give us biased results if we choose it for missing value imputation. In this connection, we will use the median to impute the missing values through the below code.

```
Data_Titanic['Age'].median(skipna=True)
```

➢ **Output: 28.0**

In the above code, **'skipna=True'** removes the null values and calculates the median on the rest data points.

- **Cabin–Missing Values:** Similar to the 'Age' variable, we need to compute the proportion of missing observation against the variable 'Cabin' using the codes given below. In the below codes 665 is the number of observation where the value of the **Cabin** variable is null.

```
round(665/len(Data_Titanic),3)
```

- **Output: 0.786**

We have realized that ~79% of records are missing; this indicates that it is probably not wise to impute the missing values and use the 'Cabin' variable for further analysis. Thus, we may ignore this variable in our model or make a separate category for passengers with the null value.

Embarked–Missing Values

Similarly, the proportion of missing values for the variable 'Embarked' has been computed as below.

```
round(2/len(Data_Titanic),3)
```

- **Output: 0.002**

There are only 2 missing values for 'Embarked,' so we can impute with the port where most people boarded (Mode) since the variable is nominal type.

Previously we have used the **matplotlib**, a data visualization library. Further, we will use **Seaborn** library, which provides a high-level interface to draw statistical graphs and charts. Before using the seaborn library, we need to install the library using the below command.

```
import seaborn as sns
```

In the next step, we can set the background color of our choice. In the below bar chart, the background is **white.** The style must be one of white, dark, white grid, dark grid, tickswhite background style for seaborn plots. Figure 7.3 exhibits the bar chart for the feature/variable 'Embarked Port' using seaborn library.

```
sns.set(style='white')
```

```
sns.countplot(x='Embarked',data=Data_Titanic,palette='Set3')
plt.show()
```

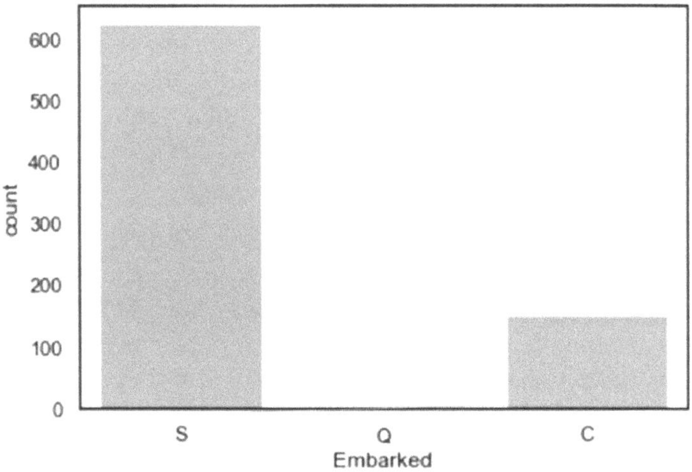

FIGURE 7.3 Bar chart for embarked port.

Further, we can show the grid lines using **white grid** and change the color combinations using 'Set1.' Please refer to Figure 7.4.

```
sns.set(style="whitegrid")
```

```
sns.countplot(x='Embarked',data=df,palette='Set1')
plt.show()
```

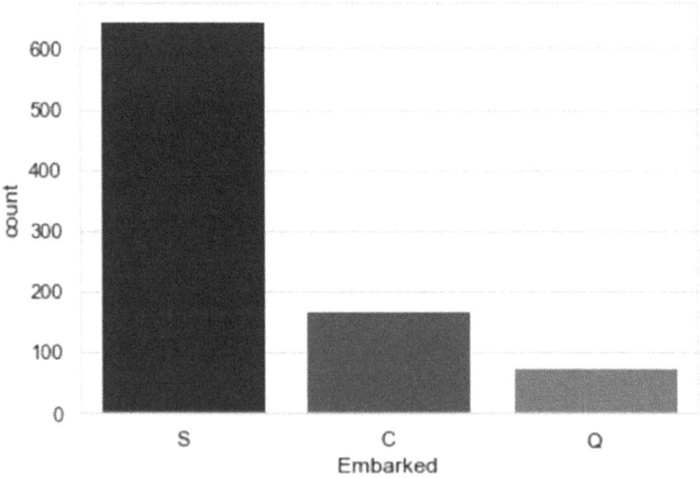

FIGURE 7.4 Bar chart for embarked port with grid lines and different color set.

Since most passengers boarded in Southampton, we will impute two Null observations with **"S."**

7.5.1.2 REPLACING THE NULL VALUES

Now, we will replace the null values with the other values as discussed above. In the below codes, the null values in the 'Age' variable have been replaced by the median age, i.e., 28 years. Similarly, null values in the 'Embarked' variable have been replaced by 'S,' which means Southampton. The mentioned code **inplace=True** depicts that the data set Data_Titanic will be overwritten with these changes (null value imputation).

```
Data_Titanic["Age"].fillna(28, inplace=True)
```

```
Data_Titanic["Embarked"].fillna("S", inplace=True)
```

7.6 REMOVING A VARIABLE FROM THE DATA SET

We have observed that the variable 'Cabin' consists of 79% missing or null values; therefore, it is recommended that this variable should be dropped from the datasets and not be used in further analysis. **'Drop'** function has been used as below to delete the column 'Cabin.'

```
Data_Titanic.drop('Cabin', axis=1, inplace=True)
```

```
Data_Titanic.isnull().sum()
```

```
PassengerId    0
Survived       0
Pclass         0
Name           0
Sex            0
Age            0
SibSp          0
Parch          0
Ticket         0
Fare           0
Embarked       0
dtype: int64
```

Here, we can observe that the null values have been replaced. Thus, there is no null value present in the variable preview of null counts.

7.7 COMPUTATION OF NEW VARIABLES

According to the data dictionary, both the variables **'SibSp'** and **'Parch'** show the number of companions. In order to avoid possible multicollinearity, we can combine the effect of these variables into one categorical feature, i.e., whether or not that individual was a solo passenger.

First, we added the two variables **'SibSp'** and **'Parch'** to get the total count of the companions with the passenger. We have stored the total number of companions in a new variable, 'Traveller_Count.' The below-mentioned code is used to calculate the total count of the companions.

```
Data_Titanic['Traveller_Count']=Data_Titanic['SibSp']+Data_Titanic['Parch']
```

Furthermore, to compute the variable **'Solo_Traveller,' numpy** library has been imported. Using the **'where'** function, a passenger is marked 'NO' in the variable 'Solo_Traveller' if the total count of companions is greater than 0.

```
import numpy as np

Data_Titanic['Solo_Traveller']=np.where(Data_Titanic['Traveller_Count']>0, 'NO', 'YES')
```

In order to use the continuous variable 'Age' as a Categorical variable, a new variable, 'Age_Group' has been computed using various conditions. This computed categorical variable has Six categories, as mentioned in the below codes.

```
Data_Titanic['Age_Group']=np.select([Data_Titanic['Age']<=15,(Data_Titanic['Age']>15) & (Data_Titanic['Age']<=25),
                (Data_Titanic['Age']>25) & (Data_Titanic['Age']<=35),(Data_Titanic['Age']>35) &
                                  (Data_Titanic['Age']<=45),
                (Data_Titanic['Age']>45) & (Data_Titanic['Age']<=55),
                (Data_Titanic['Age']>55)],['15 or Below','16-25','26-35','36-45','46-55','Above 55'],default=0)
```

The syntax mentioned below enables us to glance at the distribution of age categories in the data.

```
Data_Titanic['Age_Group'].value_counts()
```

```
26-35          346
16-25          204
36-45          115
15 or Below     81
46-55           64
Above 55        36
Name: Age_Group, dtype: int64
```

- Ascending the data view corresponding to a particular variable

In order to have a glance at data with the ascending order of age group, the following syntax may be used

```
Data_Titanic['Age_Group'].value_counts()
```

PassengerId	Survived	Pclass	Name	Sex	Age	SibSp	Parch	Ticket	Fare	Embarked	Traveller_Count	Solo_Traveller	Age_Group
814	0	3	Andersson, Miss. Ebba Iris Alfrida	female	6.0	4	2	347082	31.2750	S	6	0	15 or Below
788	0	3	Rice, Master. George Hugh	male	8.0	4	1	382652	29.1250	Q	5	0	15 or Below
44	1	2	Laroche, Miss. Simonne Marie Anne Andree	female	3.0	1	2	SC/Paris 2123	41.5792	C	3	0	15 or Below
751	1	2	Wells, Miss. Joan	female	4.0	1	1	29103	23.0000	S	2	0	15 or Below
148	0	3	Ford, Miss. Robina Maggie "Ruby"	female	9.0	2	2	W./C. 6608	34.3750	S	4	0	15 or Below

7.8 EXPLORATORY DATA ANALYSIS

Exploratory Data Analysis is a critical process of executing initial investigations on data to discover patterns, spot anomalies, test hypotheses, and check assumptions with the help of summary statistics and graphical representations. In this section, we will explore the bivariate distributions of the independent variable and dependent variable.

7.8.1 BI-VARIATE EXPLORATION OF AGE

The below-mentioned code explores the unique value of the age of the passengers.

```
Data_Titanic.Age.unique()
```

```
array([21.  , 28.  , 30.5 , 19.  , 48.  , 35.  , 18.  , 45.  , 36.  ,
       31.  , 30.  , 41.  , 38.  , 62.  , 43.  , 23.  , 25.  , 37.  ,
       22.  , 24.  , 16.  ,  2.  , 40.  , 54.  , 65.  , 11.  , 47.  ,
        3.  , 52.  , 29.  , 42.  , 55.  , 32.  , 34.  , 15.  ,  5.  ,
        1.  , 51.  ,  6.  , 27.  ,  0.75, 17.  , 33.  , 71.  ,  4.  ,
       26.  , 55.5 , 50.  ,  9.  , 60.  , 61.  , 80.  ,  7.  , 57.  ,
       39.  , 44.  , 23.5 , 34.5 , 64.  ,  8.  , 20.  ,  0.67, 63.  ,
       56.  , 32.5 ,  0.42, 46.  , 49.  , 59.  , 45.5 , 24.5 , 40.5 ,
       14.  , 20.5 , 70.5 ,  0.92, 28.5 , 66.  ])
```

The below-mentioned codes give the density plot (Figure 7.5) of age with respect to the probability of survival of the passengers.

```
plt.figure(figsize=(12,6))
sns.kdeplot(Data_Titanic["Age"][Data_Titanic.Survived == 1], color="sandybrown", shade=True)
plt.title('Density Plot of Age for Surviving Population and Deceased Population')
plt.show()
```

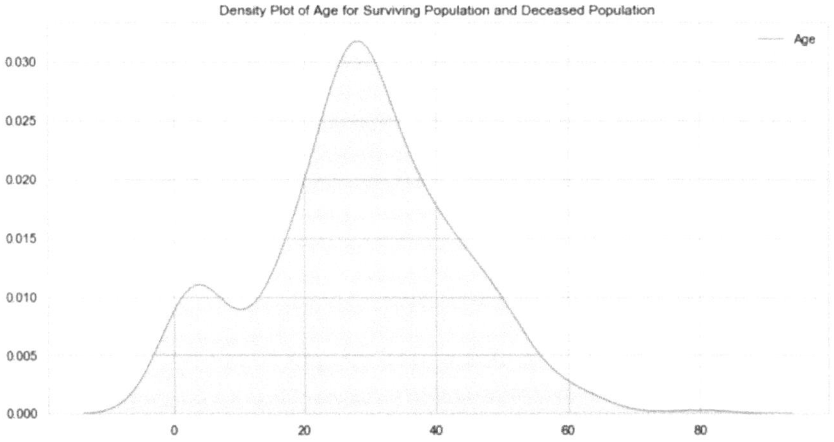

FIGURE 7.5 Density plot of age with respect to survival probability.

Using the below codes, we have plotted the age distribution of survived and died passengers in a single density plot (Figure 7.6).

```
plt.figure(figsize=(12,6))
sns.kdeplot(Data_Titanic["Age"][Data_Titanic.Survived == 1], color="sandybrown", shade=True)
sns.kdeplot(Data_Titanic["Age"][Data_Titanic.Survived == 0], color="teal", shade=True)
plt.legend(['Survived', 'Died'])
plt.title('Density Plot of Age for Surviving Population and Died Population')
plt.show()
```

FIGURE 7.6 Density plot of age for surviving population and deceased population.

The age distributions for survivors and dead passengers are pretty similar. However, it is worth noting that, of the survivors, a more significant proportion were children. This may be a piece of evidence that the passengers had made an attempt to save children by giving them the priority and provide lifeboats.

7.8.2 BI-VARIATE EXPLORATION OF FARE

```
plt.figure(figsize=(12,6))
sns.kdeplot(Data_Titanic["Fare"][Data_Titanic.Survived == 1], color="sandybrown", shade=True)
sns.kdeplot(Data_Titanic["Fare"][Data_Titanic.Survived == 0], color="teal", shade=True)
plt.legend(['Survived', 'Died'])
plt.title('Density Plot of Fare for Surviving Population and Died Population')
plt.xlim(-15,150)
plt.show()
```

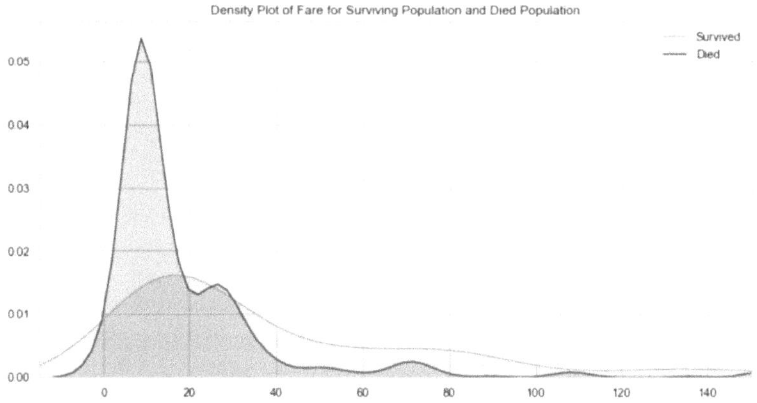

FIGURE 7.7 Density plot of fare for surviving and deceased population.

Figure 7.7 depicts that distributions are significantly different for the fares of survivors vs. deceased, this would likely be a significant predictor in our model development process. Passengers who paid lower fares appear to have been less likely to survive. This result suggests that there is probably a strong correlation between with Passenger Class and survival, which we will look at next model development section.

7.8.3 BI-VARIATE EXPLORATION OF PASSENGER CLASS

The below-mentioned codes explore the distribution of survival probabilities among the passenger classes.

```
sns.barplot('Pclass', 'Survived', data=Data_Titanic, color="lightgreen")
plt.show()
```

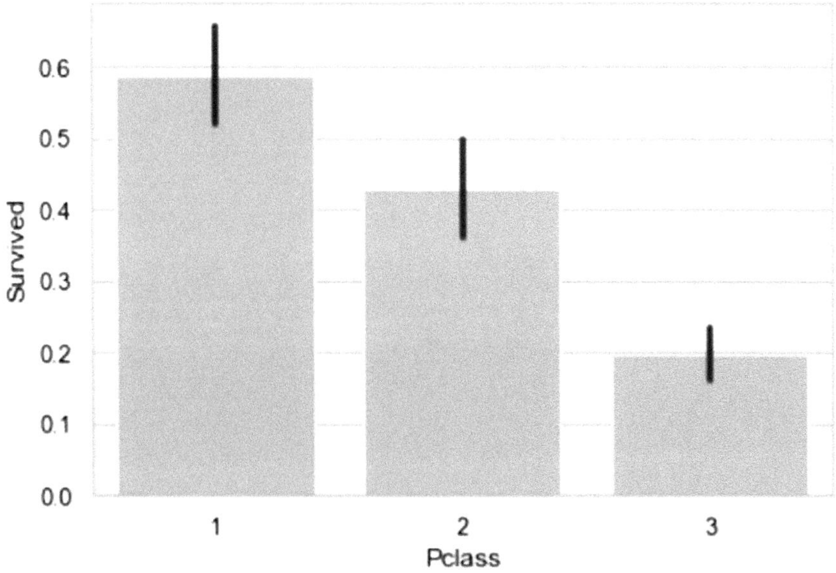

FIGURE 7.8 Survival probabilities among passenger classes.

Figure 7.8 gives the survival probabilities among all passenger classes. Inevitably, being a first-class passenger was safest and had the highest propensity of survival.

7.8.4 BI-VARIATE EXPLORATION OF EMBARKED PORT

The below-mentioned codes exhibit the distribution of survival probabilities among the Embarkation ports (Figure 7.9).

```
sns.barplot('Embarked', 'Survived', data=Data_Titanic, color="aqua")
plt.show()
```

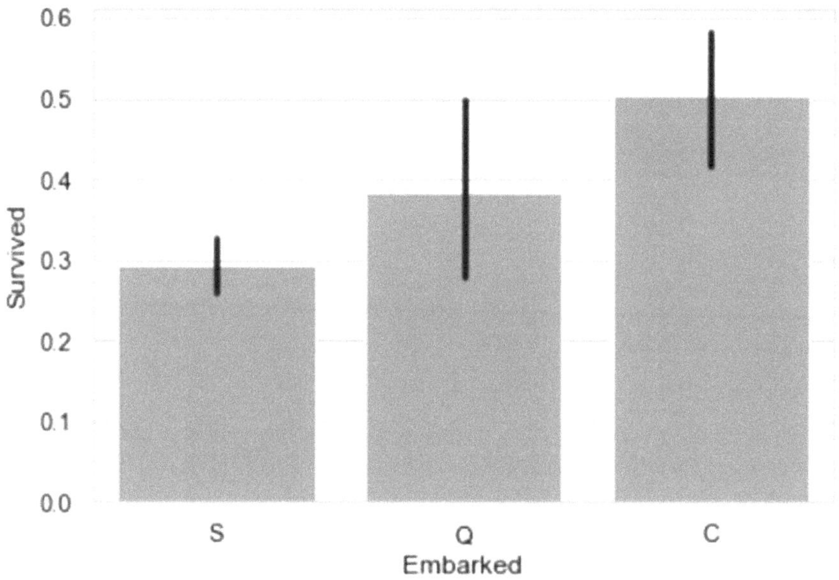

FIGURE 7.9 Survival probabilities among embarked ports.

Exploring the above distribution, we can state that the passengers who boarded in Cherbourg, France, appear to have the highest survival rate. Passengers who boarded in Southampton were marginally less likely to survive than those who boarded in Queenstown. This is probably related to passenger class, or maybe even the order of room assignments (e.g., maybe earlier passengers were more likely to have rooms closer to deck).

7.8.5 BI-VARIATE EXPLORATION OF SOLO TRAVELER

```
sns.barplot('Solo_Traveller', 'Survived', data=Data_Titanic, color="cadetblue")
plt.show()
```

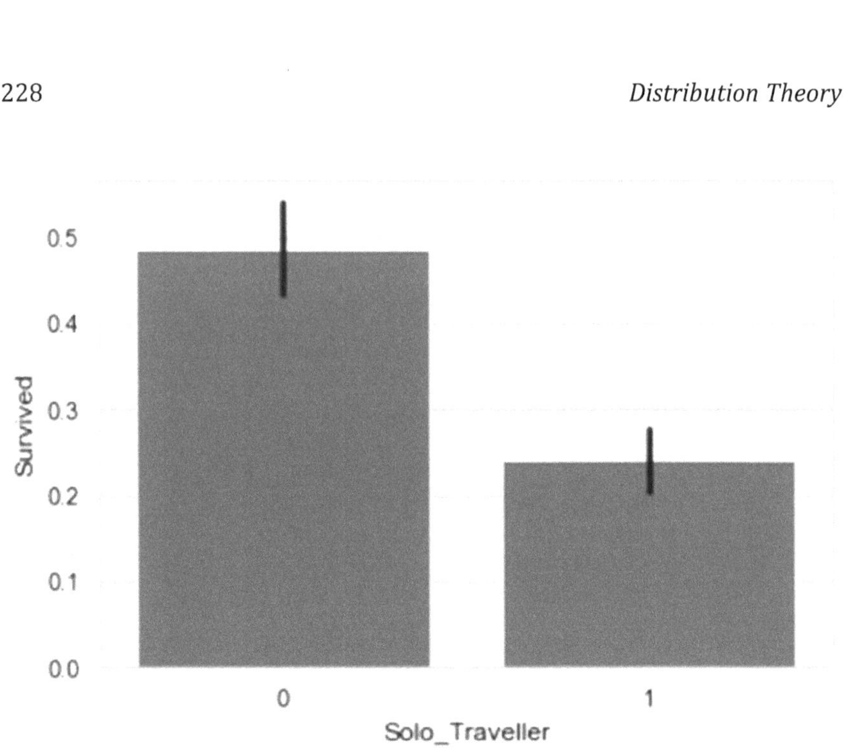

FIGURE 7.10 Survival probabilities v/s. solo passengers.

Figure 7.10 depicts that individuals traveling without families were more likely to die in the disaster than those with families. For the Given incident, there is a possibility that individuals traveling alone were likely male.

7.8.6 *BI-VARIATE EXPLORATION OF GENDER VARIABLE*

```
sns.barplot('Sex', 'Survived', data=Data_Titanic, color="darkolivegreen")
plt.show()
```

The above-mentioned finding in Figure 7.11 exhibits a pronounced difference. Clearly, female passengers were given preference to get into the lifeboat first, increasing their chances of survival.

7.9 CREATING DUMMY VARIABLES

In order to use the categorical variables in the model, we need to create dummy variables for all the categorical variables, which seem essential for model building. We can refer to the codes mentioned below to create dummy variables for the categorical predictors/independent variables **'Pclass,'** which refers to the passenger class.

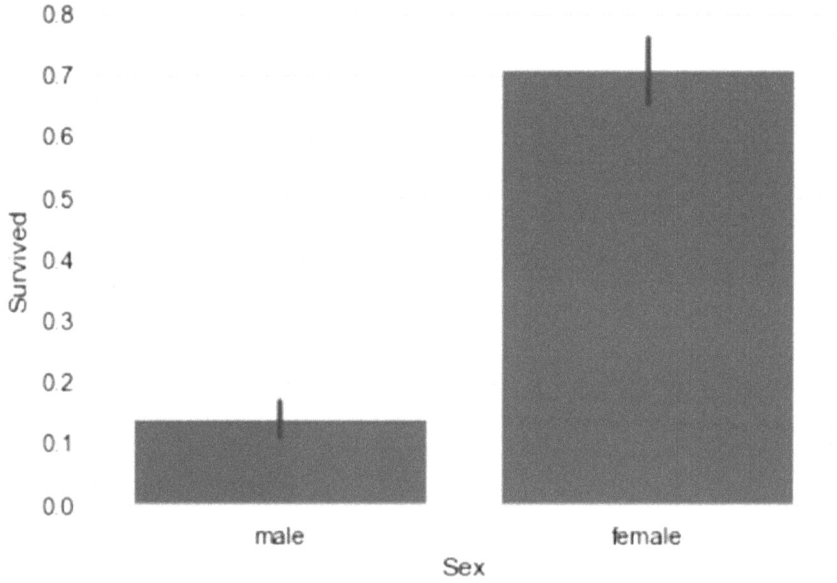

FIGURE 7.11 Survival probabilities v/s gender of traveler.

```
Data_Titanic1 = pd.get_dummies(Data_Titanic, columns=["Pclass"])
```

In order to create multiple dummy variables in one step, the following syntax may be utilized.

```
Data_Titanic2 = pd.get_dummies(Data_Titanic1,
columns=["Embarked","Sex","Age_Group","Solo_Traveller"])
```

After creating dummy variables for all desired categorical variables, the new dataset is stored as **Data_Titanic2.** The below preview of the data 'Data_Titanic2' exhibits the dummy variables for the categorical variables **'Embarked'** and **'Sex.'** The use of dummy variables will be discussed in the model development section.

```
Data_Titanic2
```

PassengerId	Survived	Name	Age	SibSp	Parch	Ticket	Fare	Traveller_Count	...	Embarked_Q	Embarked_S	Sex_female	Sex_male
38	0	Cann, Mr Ernest Charles	21.0	0	0	A./5 2152	8.0500	0	...	0	1	0	1
236	0	Harknett, Miss. Alice Phoebe	28.0	0	0	W./C. 6609	7.5500	0	...	0	1	1	0
73	0	Hood, Mr. Ambrose Jr	21.0	0	0	S.O.C. 14879	73.5000	0	...	0	1	0	1

After the variable's exploration, we have an idea about the importance of predictors for model building. Therefore, the unimportant and unnecessary variables can be dropped further as per the below syntax.

```
Data_Titanic2.drop(['SibSp','Parch','PassengerId','Name','Ticket'], axis=1, inplace=True)
```

7.10 EXPORTING THE DATA IN .CSV FILE

Once the data cleaning and preparation is completed, we should export the data in desired file format to save the processing time while using the exact data for analysis or model building in the future. The below syntax can be used to export the prepared data "Data_Titanic2" into a .csv file with the name "Data_Titanic.csv." Users may export the data in other formats of files like .txt .xlsx, etc.

```
Data_Titanic2.to_csv("C:\\Users\\Desktop\\Py\\Titan_Data\\Data_Titanic.csv")
```

7.11 MODEL DEVELOPMENT

Further, for model building, independent variables (predictors) need to be stored in an array. Therefore, we have stored all the predictors in an array **cols** as mentioned in the below python codes.

```
cols=["Age_Group_16-25","Age_Group_26-35", "Age_Group_36-45","Age_Group_46-55",
      "Age_Group_Above 55","Solo_Traveller_YES","Fare","Traveller_Count",
      "Pclass_1", "Pclass_2","Embarked_C","Embarked_S","Sex_male"]
```

In order to get the list of all variables which we will supply as independent variables in the model, we need to print the **cols** as mentioned below. The continuous independent variables are supplied as it is; however, we need to supply all the categories for categorical variables except the one we consider as the reference category. For example, the dummy variable

'Sex_male' will be supplied as a predictor to the model, and 'Sex_female' will be the reference category, which means the regression coefficients will be displayed for Male passengers with respect to Female passengers.

```
cols
```

```
['Age_Group_16-25',
 'Age_Group_26-35',
 'Age_Group_36-45',
 'Age_Group_46-55',
 'Age_Group_Above 55',
 'Solo_Traveller_YES',
 'Fare',
 'Traveller_Count',
 'Pclass_1',
 'Pclass_2',
 'Embarked_C',
 'Embarked_S',
 'Sex_male']
```

Having a glance at all the variable names stored in **cols** enables us to cross-validate if we have taken all the important predictors for the further model development process.

Further, we will split the data set into two data frames, X and Y, as given in the codes mentioned below. Here, X contains all the desired independent variables which are suitable for model building. On the other side, Y contains only one column/dependent/target variable 'Survived,' which we need to predict.

```
X=Data_Titanic2[cols]
```

```
Y=Data_Titanic2['Survived']
```

7.11.1 LOGISTIC REGRESSION MODEL

We will be exploring the logistic regression model for this binary classification problem. In order to perform the initial model exploration, we need to have the regression coefficients of the logistic regression. Therefore, the below codes for the logistic regression model have been utilized using **statsmodels.api,** and **scipy** libraries in order to get the regression coefficients. Logistic regression model using **sklearn** library will be discussed later when we need to fit the model without regression coefficients for further prediction. The below syntax can be used to fit the logistic regression model with all the desired predictors.

```
import statsmodels.api as sm
from scipy import stats
stats.chisqprob = lambda chisq, df: stats.chi2.sf(chisq, df)
logit_model=sm.Logit(Y,X)
result=logit_model.fit()
print(result.summary())
```

```
Optimization terminated successfully.
         Current function value: 0.404359
         Iterations 7
                           Logit Regression Results
==============================================================================
Dep. Variable:               Survived   No. Observations:                  846
Model:                          Logit   Df Residuals:                      833
Method:                           MLE   Df Model:                           12
Date:                Sun, 30 May 2021   Pseudo R-squ.:                  0.3679
Time:                        20:58:27   Log-Likelihood:                -342.09
converged:                       True   LL-Null:                       -541.22
Covariance Type:            nonrobust   LLR p-value:                 8.808e-78

================================================================================
                      coef    std err          z      P>|z|      [0.025      0.975]
--------------------------------------------------------------------------------
Age_Group_16-25    -0.9998      0.334     -2.990      0.003      -1.655      -0.344
Age_Group_26-35    -0.6367      0.296     -2.154      0.031      -1.216      -0.057
Age_Group_36-45    -0.7937      0.381     -2.081      0.037      -1.541      -0.046
Age_Group_46-55    -0.9396      0.439     -2.141      0.032      -1.800      -0.079
Age_Group_Above 55 -1.7549      0.618     -2.840      0.005      -2.966      -0.544
Solo_Traveller_YES -0.2030      0.258     -0.788      0.431      -0.708       0.302
Fare               -0.0007      0.003     -0.231      0.818      -0.007       0.006
Traveller_Count    -0.1673      0.081     -2.060      0.039      -0.326      -0.008
Pclass_1            2.2867      0.351      6.516      0.000       1.599       2.975
Pclass_2            1.5260      0.258      5.917      0.000       1.021       2.031
Embarked_C          1.2433      0.363      3.429      0.001       0.533       1.954
Embarked_S          0.9280      0.328      2.832      0.005       0.286       1.570
Sex_male           -2.8107      0.224    -12.527      0.000      -3.250      -2.371
================================================================================
```

Except **'Solo_Traveller'** and **'Fare,'** nearly all the independent variables are significant at the 5% level of significance. Therefore, we will rerun the model without **'Solo_Traveller'** and **'Fare.'** We removed these insignificant variables one by one in each iteration and found no significant effect on the regression results. Thus, these insignificant input variables were finally removed from the regression model and found the final regression coefficients. First of all, we need to store the remaining significant variables again in an array as mentioned in the codes mentioned below.

```
cols=["Age_Group_16-25","Age_Group_26-35", "Age_Group_36-45","Age_Group_46-55",
      "Age_Group_Above 55","Traveller_Count",
      "Pclass_1", "Pclass_2","Embarked_C","Embarked_S","Sex_male"]
```

```
cols
```

```
['Age_Group_16-25',
 'Age_Group_26-35',
 'Age_Group_36-45',
 'Age_Group_46-55',
 'Age_Group_Above 55',
 'Traveller_Count',
 'Pclass_1',
 'Pclass_2',
 'Embarked_C',
 'Embarked_S',
 'Sex_male']
```

In the preview of **cols,** we can see that the insignificant variables have been removed from the list of independent variables. Further, we will create the data frame X again using the below code for all the significant independent variables. The Y data frame, which contains the dependent variable, will remain the same as we used above.

```
X=Data_Titanic2[cols]
```

Using the below codes, we again fit the logistic regression model to find out the new regression coefficients.

```
import statsmodels.api as sm
from scipy import stats
stats.chisqprob = lambda chisq, df: stats.chi2.sf(chisq, df)
logit_model=sm.Logit(Y,X)
result=logit_model.fit()
print(result.summary())
```

The above results show a slight improvement in the p values for the selected independent variables. We further trained the logistic regression for prediction on whole data using the **sklearn** library. Referring to the below codes, LogisticRegression has been imported from the library **sklearn.linear_model.** After fitting the logistic regression model (logreg), we computed the model's accuracy using the **'.score'** function and rounded it off to the two decimal place.

```
Optimization terminated successfully.
         Current function value: 0.404735
         Iterations 7
                           Logit Regression Results
==============================================================================
Dep. Variable:               Survived   No. Observations:                  846
Model:                          Logit   Df Residuals:                      835
Method:                           MLE   Df Model:                           10
Date:                Sun, 06 Jun 2021   Pseudo R-squ.:                  0.3673
Time:                        19:11:26   Log-Likelihood:                -342.41
converged:                       True   LL-Null:                       -541.22
Covariance Type:            nonrobust   LLR p-value:                 3.005e-79

=======================================================================================
                          coef    std err          z      P>|z|      [0.025      0.975]
---------------------------------------------------------------------------------------
Age_Group_16-25        -1.0826      0.317     -3.418      0.001      -1.704      -0.462
Age_Group_26-35        -0.7414      0.265     -2.801      0.005      -1.260      -0.223
Age_Group_36-45        -0.8874      0.363     -2.443      0.015      -1.599      -0.175
Age_Group_46-55        -1.0115      0.428     -2.363      0.018      -1.850      -0.173
Age_Group_Above 55     -1.8738      0.591     -3.172      0.002      -3.032      -0.716
Traveller_Count        -0.1401      0.065     -2.142      0.032      -0.268      -0.012
Pclass_1                2.2760      0.278      8.174      0.000       1.730       2.822
Pclass_2                1.5339      0.253      6.062      0.000       1.038       2.030
Embarked_C              1.1949      0.355      3.364      0.001       0.499       1.891
Embarked_S              0.8659      0.316      2.741      0.006       0.247       1.485
Sex_male               -2.8416      0.221    -12.841      0.000      -3.275      -2.408
=======================================================================================
```

```
from sklearn.linear_model import LogisticRegression

logreg = LogisticRegression()
logreg.fit(X, Y)

round(logreg.score(X, Y),2)

import warnings
warnings.filterwarnings("ignore")
```

➢ **Output: 0.83**

The above result shows that the logistic regression is fitted on whole data with 83% accuracy of prediction. The additional code for removing the error in the output display is employed using the **warnings** library.

7.11.2 VISUALIZING THE MODEL'S ROC CURVE AND AUC

The receiver operating characteristic (ROC) curve is a performance evaluation tool for a predictive model at variously defined thresholds. In a ROC curve, the true positive (TP) rate (Sensitivity) is plotted in function of the false positive (FP) rate (1-Specificity) for different cut-off points.

Each point on the ROC curve represents a sensitivity/specificity pair corresponding to a particular decision threshold.

The below-mentioned codes are utilized to display the ROC curve (Figure 7.12) for the logistic regression model trained on the complete Titanic data set. We have imported **roc_curve** and **auc** from the **sklearn.metrics** as mentioned in the codes. The codes involve **matplotlib.pyplot** library as mentioned in the previously mentioned graphs.

```
from sklearn.metrics import roc_curve, auc

plt.figure(figsize=(9,7))
plt.plot(FPR, TPR, label='ROC curve (area = %0.2f)' % roc_auc)
plt.plot([0, 1], [0, 1], 'k--')
plt.xlim([-0.05, 1.0])
plt.ylim([0.0, 1.05])
plt.xlabel('False Positive Rate')
plt.ylabel('True Positive Rate')
plt.title('ROC Curve')
plt.legend(loc="lower right")
plt.show()
```

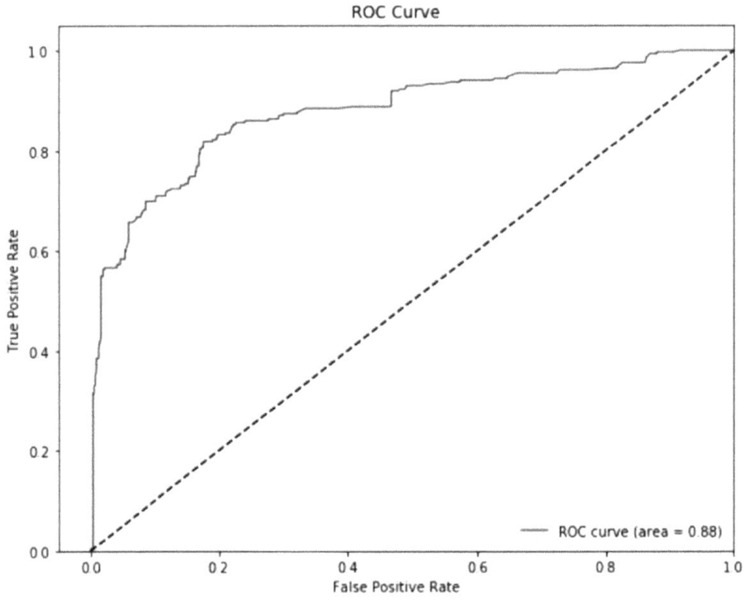

FIGURE 7.12 ROC curve of the model trained on complete data set.

The area under curve (AUC) is the area enclosed by the ROC curve. A perfect classifier has AUC = 1, and a completely random classifier has AUC = 0.5. Usually, a good model scores somewhere in between. The range of possible AUC values is [0, 1]. The below-mentioned codes are utilized to obtain the value of AUC.

```
FPR, TPR, _ = roc_curve(Y, logreg.predict_proba(X)[:,1])

roc_auc = auc(FPR, TPR)
print ('ROC AUC: %0.2f' % roc_auc )
```

➢ **Output: ROC AUC: 0.88**

It is noted in the above output that the value of AUC is 0.88; it means there is ~88% chance that the model will distinguish between positive class (passenger survived) and negative class (passenger not survived).

7.11.3 VALIDATION OF THE MODEL USING TRAIN AND TEST PARTITIONS

The model developed on complete data provides good accuracy; however, it is necessary to test the model on a partition of the complete data that is not used while training the model. It is a general convention to take the 70:30 or 80:20 to split the data into train and test partition. However, one may take other ratios as well to split the data into training and testing sets.

In order to split the data into train and test partitions, we have imported **train_test_split** function from the **sklearn.model_selection.** In this classification problem, the complete data have been partitioned into train and test data sets in the ratio of 70:30, which means that 70% of data points (passengers) will be considered for the model training, and the remaining 30% will be utilized to test the model predictions.

The codes mentioned below are used to split the complete data (Data_Titanic2) into train and test data sets with 70:30.

```
from sklearn.model_selection import train_test_split
train, test = train_test_split(Data_Titanic2, test_size=0.3)
```

Similar to the previous section, X2 and Y2 data sets are stored containing the independent and dependent variables, respectively, for the training partition. After that, we have re-fitted/re-trained the logistic model for the training data set.

```
X2=train[cols]
Y2=train['Survived']
```

```
logreg = LogisticRegression()
logreg.fit(X2, Y2)
round(logreg.score(X, Y),2)
```

➢ **Output: 0.84**

The score for the new training sample (84% of original) is very close to the original performance, which is good. Let us assess how well it scores on the 30% hold-out sample (testing partition).

The below codes show that X3_test and Y3_test data sets are stored containing the independent and dependent variables, respectively, for the testing partition. Further, the prediction for the testing data is stored into a data frame **Y3test_pred** using the **logreg.predict** function on the LR model (logreg), which was trained on the training dataset.

```
X3_test = test[cols]
Y3_test = test['Survived']
Y3test_pred = logreg.predict(X3_test)
```

7.11.4 CONFUSION MATRIX AND MODEL PERFORMANCE

The required libraries and functions are imported to evaluate the model accuracy, confusion matrix, and other model performance parameters as given in the below codes.

```
from sklearn import metrics
from sklearn.metrics import confusion_matrix
from sklearn.metrics import classification_report
```

The accuracy of the trained model on testing data can be found using the codes mentioned below. The output shows that the model accuracy is 86% on the testing dataset. This indicates that the trained LR model is performing well on the testing dataset also.

```
print("Accuracy:",metrics.accuracy_score(Y3_test, Y3test_pred))

Accuracy: 0.8622047244094488
```

The below-mentioned codes are employed to find out the confusion matrix, which is a matrix that exhibits the TP-63, true negative (TN)-156, FP-13, false negative (FN)-22.

```
print(confusion_matrix(Y3_test, Y3test_pred))

[[156  13]
 [ 22  63]]
```

The below code is used to obtain the classification report, which consists of precision, recall, and f1-score on the testing data set here.

```
print(classification_report(Y3_test, Y3test_pred))

              precision    recall  f1-score   support

           0       0.88      0.92      0.90       169
           1       0.83      0.74      0.78        85

    accuracy                           0.86       254
   macro avg       0.85      0.83      0.84       254
weighted avg       0.86      0.86      0.86       254
```

In the output mentioned above, **recall** value (0.92) corresponding to "0" class (not survived) suggests that out of total actual dead passengers, the model can predict 92% dead passengers correctly. The recall value of 0 class is also called **specificity**. Similarly, recall value (0.74) corresponding to "1" class (survived) suggests that out of total actual survived passengers, the model is able to predict 74% of survived passengers correctly. This value is also known as **sensitivity**.

The **precision** value for class 1 (survived) 0.83 depicts that out of total passengers predicted as survived by the model, 83% actually survived. Similarly, the precision value for class 0 (not survived) 0.88 depicts that out of total passengers who are predicted as not survived by the model, 88% are actually dead. Further, the f1-score is the harmonic mean of the precision and recall corresponding to each class of dependent variable. After glancing at the f1-score values for each prediction class, we can state that the model performs very well on the testing data set.

7.11.5 ASSESSING THE MODEL PERFORMANCE BASED ON ROC/ AUC (TESTING DATA)

As we discussed earlier, the ROC curve has been obtained as below to access the model performance. The ROC curve for the testing data (Figure 7.13) looks similar to the one which was obtained for complete data using the LR model.

```
plt.figure(figsize=(9,7))
plt.plot(FPR, TPR, label='ROC curve (area = %0.2f)' % roc_auc)
plt.plot([0, 1], [0, 1], 'k--')
plt.xlim([-0.05, 1.0])
plt.ylim([0.0, 1.05])
plt.xlabel('False Positive Rate')
plt.ylabel('True Positive Rate')
plt.title('ROC Curve')
plt.legend(loc="lower right")
plt.show()
```

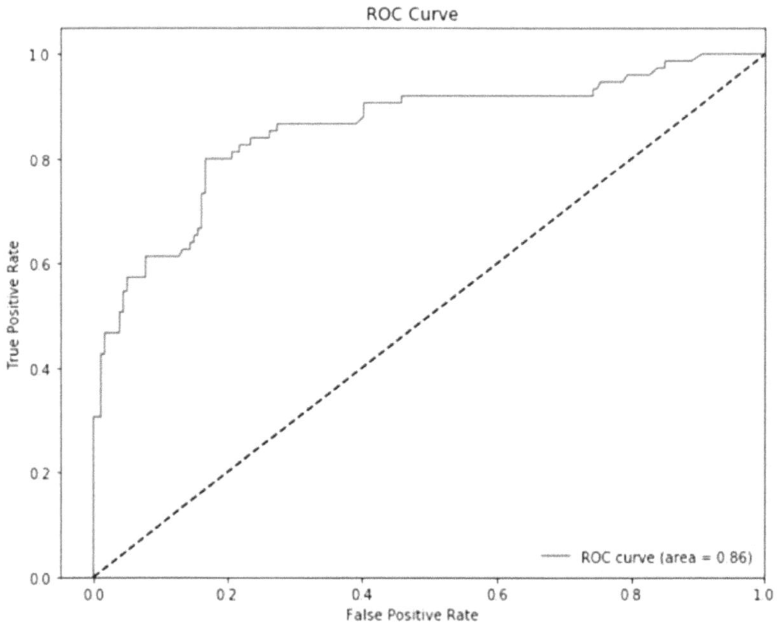

FIGURE 7.13 ROC curve of LR model on testing data set.

```
FPR, TPR, _ = roc_curve(Y3_test, logreg.predict_proba(X3_test)[:,1])

roc_auc = auc(FPR, TPR)
print ('ROC AUC: %0.2f' % roc_auc )
```

➢ **Output: ROC AUC: 0.86**

The AUC value for the testing data set 0.86 is pretty close to the AUC value (0.88) obtained after building the LR model on the complete dataset. Therefore, we can state that the trained model predicts the survival of the passenger quite accurate and gives the stable results on testing data as well.

7.11.6 SAVING THE TRAINED MODEL INTO PICKLE FILE

Once a well-tuned model is trained with the desired accuracy, we should save the model for future use or prediction. In order to serve this purpose first, we need to import the pickle library using the below codes.

```
import pickle
```

In order to save the .pkl file to save the model, the following codes can be referred. A .pkl file for our trained model logreg is saved at our desired location in the given codes. The name of the saved model file is **Titanic_Model.pkl** which can be retrieved in the future to reuse this model.

```
with open("C:\\Users\\Desktop\\Py\\Titan_Data\\Titanic_Model.pkl", 'wb') as f:
        pickle.dump(logreg, f)
```

7.11.7 LOADING THE SAVED MODEL

If a .pkl file is saved for a model, the same can be retrieved for further use. The below-mentioned codes can be utilized to load the model for further prediction.

```
with open("C:\\Users\\Desktop\\Py\\Titan_Data\\Titanic_Model.pkl", 'rb') as f:
    logreg = pickle.load(f)
```

KEYWORDS

- **classification**
- **confusion matrix**
- **data exploration**
- **data visualization**
- **model development**
- **python**
- **titanic data set**

REFERENCES

Beazley, J., (2009). *The Python Essential Reference* (4th edn.). Addison-Wesley Professional, 2009.

Bernard, J., (2008). Use python for scientific computing. *Linux Journal, 175*(7), 2008.

Lutz, M., (2013). *Learning Python* (5th edn.). O'Reilly Media.

Wolfram Research, (2018). *Sample Data: Titanic Survival.* The Wolfram Data Repository.

Index

D

E

L

M

N

R

S